*Energy Saving Tips
That Will Save You MONEY!*

I0487004

Copyright © 2008
By American Environmental
East Tawas, Michigan 48730

Published By:
American Environmental
East Tawas, MI. 48730

This book was created by American
Environmental as a reference source of
information for solutions to lowering your
energy consumption and expenses with FREE or
Low-Cost solutions. American Environmental.

ISBN-10: 1-440475-03-2

Energy Saving Tips That Will Save You MONEY!

ACKNOWLEDGEMENTS

"Energy Saving Tips That Will Save You MONEY" was developed as a resource for the Eco-Environmental Corporation to help people discover unknown expenses in their electric and gas consumption. By reducing these expenses we can not only help save the environment from the harmful byproducts of electric and gas use, but we can also save money at the same time!

This book contains hundreds of FREE as well as Low-Cost solutions for saving money that just about anyone can do themselves.

These techniques address two of the biggest issues facing our society and country today, preserving the Earth's Ecological state while providing renewable energies for citizens, and cutting waste and consumption of these energy sources, in turn, saving millions of dollars per year.

"Energy Saving Tips That Will Save You MONEY" is a straight forward, no non-sense book which cuts right to the chase, and shows you what you can do NOW to save, and what you can do in the future to save even more!

Energy Saving Tips That Will Save You MONEY!

TABLE OF CONTENTS

PART 1 – ELECTRICITY
1.1 Your Air Conditioner and You   9
1.2 Appliances and Home Electronics   15
1.3 Space Heaters / Coolers   24
1.4 Lighting and Maintenance   29
1.5 Lighting Controls   31
1.6 Your Home Computer   36
1.7 Reducing Your Electrical Consumption   41

PART 2 – NATURAL GAS / PROPANE
2.1 Insulation In Your Home   62
2.2 Winter Storm Tips   67
2.3 Reducing You Gas / Propane Consumption 71

PART 3 – YOUR WATER HEATER
3.1 Efficient Water Heating   72
3.2 Drain Recovery Systems   74
3.3 Heat Traps for Your Tank   77
3.4 Timers and Peak Times   79
3.5 Insulating Your Tank   81
3.6 Lowering Your Tanks Energy Consumption 85
3.7 Reducing Hot Water Use   88

PART 4 – HUNDREDS OF ENERGY SAVING TIPS
4.1 How Much Energy Am I Using?   97
4.2 Electric Saving Tips   108
4.3 General Money Saving Tips   116
4.4 Winter Savings   165

GLOSSARY   170

Energy Saving Tips That Will Save You MONEY!

INTRODUCTION

This information and this book was collected from well known resources both in print, and over the internet as well as handy-man tips that have been handed down.

For the first time, information from many of the leading energy conservation groups, grass root foundations, ecological preservation groups, as well as government and business organizations has been researched, compiled and put into an easy to read, easy to understand book format.

Although much of this information is not considered to be new, many of the tips and suggestions listed in this book still go unused today for no apparent reason other than a lack of knowledge.

What this book is attempting to do is bring to light as many of the easy to do solutions for the average person, so together we can all work toward a future which consumes less energy, saving people money and the Earth from the damaging effects of over consumption of energy.

With the economy the way it is, and the cost of energy, including electricity and natural gas as well as propane, increasing weekly, there are many reasons why the information in this book should be used, and used today!

By implementing the principles discussed in this book, we can promise you will save money. Obviously we can't promise how much, because that all depends on which principles you apply, how many you implement, and how long you stick with the program.

The US average energy consumption for typical family home is around $2350 per month, with nearly 75% of that cost coming from home appliances and electronics. Energy efficiency must be a life style, not a temporary fad.

Energy Saving Tips That Will Save You MONEY!

PART 1

ELECTRICITY

1.1

YOUR AIR CONDITIONER AND YOU

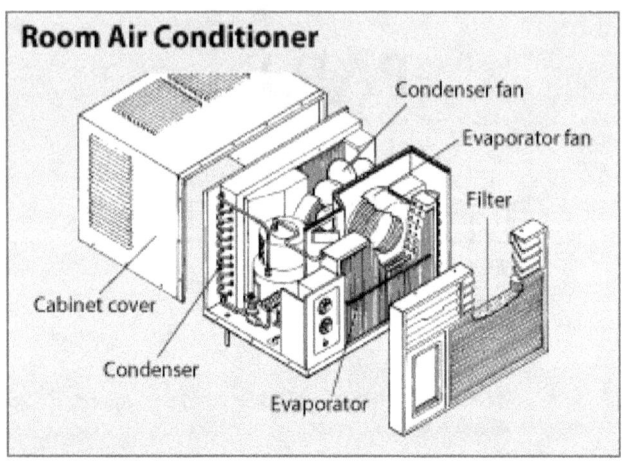

Room Air Conditioner

Condenser fan

Evaporator fan

Filter

Cabinet cover

Condenser

Evaporator

A room air conditioner features a condenser on the end that faces the outside and a condenser fan behind it that blows air through it, helping to remove the heat from the condenser. On the end facing the room is the evaporator, with an evaporator fan behind that to push the cool air into the room. The filter is mounted in the front grill.

Room air conditioners, sometimes referred to as window air conditioners, cool rooms rather than the entire home or business. If they provide cooling only where they're needed, room air conditioners are less expensive to operate than central units, even though their efficiency is generally lower than that of central air conditioners.

Smaller room air conditioners (i.e., those drawing less than 7.5 amps of electricity) can be plugged into any 15- or 20-amp, 115-volt household circuit that is not shared with any other major appliances. Larger room air conditioners (i.e., those drawing more than 7.5 amps) need their own dedicated 115-volt circuit. The largest models require a dedicated 230-volt circuit.

Energy Efficiency of Room Air Conditioners
A room air conditioner's efficiency is measured by the energy efficiency ratio (EER). The EER is the ratio of the cooling capacity (in British thermal units [Btu] per hour) to the power input (in watts). The higher the EER rating, the more efficient the air conditioner. National appliance standards require room air conditioners built after January 1, 1990, to have an energy efficiency ratio (EER) of 8.0 or greater.

The Association of Home Appliance Manufacturers reports that the average EER of room air conditioners rose 47% from 1972 to 1991. If you own a 1970's-vintage room air conditioner with an EER of 5 and you replace it with a new one with an EER of 10, you will cut your air conditioning energy costs in half.

When buying a new room air conditioner, look for units with an EER of 10.0 or above. Check the EnergyGuide label for the unit, and also look for room air conditioners with the ENERGY STAR® label.

Sizing and Selecting a Room Air Conditioner
The required cooling capacity for a room air conditioner depends on the size of the room being cooled: Room air conditioners generally have cooling capacities that range from 5,500 Btu per hour to 14,000 Btu per hour. A common rating term for air conditioning size is the "ton," which is 12,000 Btu per hour.

Proper sizing is very important for efficient air conditioning. A bigger unit is not necessarily better because a unit that is too large will not cool an area uniformly. A small unit running for an extended period operates more efficiently and is more effective at dehumidifying than a large unit that cycles on and off too frequently.

Based on size alone, an air conditioner generally needs 20 Btu for each square foot of living space. Other important factors to consider when selecting an air conditioner are room height, local climate, shading, and window size.

Verify that your home's electrical system can meet the unit's power requirements. Room units operate on 115-volt or 230-volt circuits. The standard household receptacle is a connection for a 115-volt branch circuit. Large room units rated at 115 volts may require a dedicated circuit and room units rated at 230 volts may require a special circuit.

If you are mounting your air conditioner near the corner of a room, look for a unit that can direct its airflow in the desired direction for your room layout. If you need to mount the air conditioner at the narrow end of a long room, then look for a fan control known as "Power Thrust" or "Super Thrust" that sends the cooled air farther into the room.

Other features to look for:

- A filter that slides out easily for regular cleaning
- Logically arranged controls
- A digital readout for the thermostat setting, and
- A built-in timer.

Installing and Operating Your Room Air Conditioner
A little planning before installing your air conditioner will save you energy and money. The unit should be level when installed, so that the inside drainage system and other mechanisms operate efficiently. If possible, install the unit in a shaded spot on your home's

north or east side. Direct sunshine on the unit's outdoor heat exchanger decreases efficiency by as much as 10%. You can plant trees and shrubs to shade the air conditioner, but do not block the airflow.

Don't place lamps or televisions near your air-conditioner's thermostat. The thermostat senses heat from these appliances, which can cause the air conditioner to run longer than necessary.

Set your air conditioner's thermostat as high as is comfortably possible in the summer. The less difference between the indoor and outdoor temperatures, the lower your overall cooling bill will be. Don't set your thermostat at a colder setting than normal when you turn on your air conditioner; it will not cool your home any faster and could result in excessive cooling and unnecessary expense.

Set the fan speed on high, except on very humid days. When humidity is high, set the fan speed on low for more comfort. The low speed on humid days will cool your home better and will remove more moisture from the air because of slower air movement through the cooling equipment. Consider using an interior fan in conjunction with your window air conditioner to spread the cooled air more effectively through your home without greatly increasing electricity use.

1.2
APPLIANCES AND HOME ELECTRONICS

Appliances and Home Electronics

If you live in a typical U.S. home, your appliances and home electronics are responsible for about 20% of your energy bills. These appliances and electronics include the following:

- Clothes washers and dryers
- Computers
- Dishwashers
- Home audio equipment
- Refrigerator and freezers
- Room air conditioners
- Televisions, DVD players, and VCRs
- Water heaters

A Dryer That Plays Nicely

The big thing to look for in a new dryer is whether it has a moisture-sensing feature. A moisture sensor allows the dryer to automatically turn itself off when the clothes reach a specified level of dryness.

Next-best would be a dryer with an automatic cut-off that turns itself off by sensing the temperature of the air being exhausted. A dryer that only lets you set a timer will cost you 10-

15% more in energy usage than the ones with automatic cut-offs. Buy a good enough new dryer and your energy utility might provide you with a rebate. Call them up and ask them about their dryer rebates.

Finding the Best Washer
An "H-axis" washer is the most efficient kind. Most of these are front-loaders. They use less water (therefore less hot water, therefore less energy to heat the water). A front-loader washer has the added bonus of spinning faster, which gets rid of more moisture so your dryer doesn't have to work so hard. It takes more energy to cook wet clothes dry than spin out the moisture. Some people will tell you a front-loader washer get clothes cleaner. I have no opinion on this.

You want a washer that lets you choose warm and cold wash settings and cold rinse settings. Also look for a washer that lets you lower the water level for smaller loads.

A suds-saver setting is a plus, too.
Check with your water and electricity utilities to see if they have rebates for any energy-efficient washing machines.

By the way, EnergyGuide labels are great and all, but they aren't required to tell you what you really need to know. They give you the estimated energy use for 416 loads of laundry per year for a washer. However, some washers hold larger loads, some smaller. What you need to know is the energy factor, which combines the washer capacity and energy consumption per cycle. Some manufacturers will provide this on request.

If this is all making you worry that your ninth grade algebra teacher is going to give you a pop quiz, take heart. You can't go wrong if you get a washer with an EnergyStar® rating.

You want to install the washer as close to the water heater as possible and insulate the water pipes between the washer and water heater. Installation can affect your energy costs, too.

Dryer Power-Saving
Use your dryer efficiently to save energy when drying laundry.

The best advice for using your dryer is don't. Air dry some or all of your laundry. For those of you who don't have a clothesline in the backyard--or a backyard, for that matter--you can set up a stand-alone clothes rack and dry shirts, etc. on it. You get a bonus of humidifying the air in your home without paying to run a humidifier.

All right, you must use your dryer for many of your clothes, but you can still take some energy-saving steps:

Clean the dryer filter before each use. (If you live in my apartment building, clean the dryer lint trap before you leave the laundry room, too.)

Dry full loads, but don't overfill. Lucky you if you have matching washer dryer sets since they take the same sized loads.

Dry similar types of clothes together-- lightweight fabrics with lightweight fabrics; thick stuff with thick stuff. In other words, don't mix your lacy underthings with your bath towels in the dryer.

Don't aim for bone dry. You'll kill your clothes sooner. Use the automatic moisture-sensing shut-off if your dryer has one instead of the timer.

If you take out clothes of the dryer while still slightly damp, you won't need to do as much ironing. That is, if you hang them up right away after removing them from the dryer

Don't toss wet items into a partially finished load.

Put two loads in a row in the dryer to make the most of the heat generated the first time around. Heck, go wild and do three dryer loads in a row!

Dryers (and washers to a lesser extent) like to live in warm spaces, so they will cost you less on your energy bill if you place them in an insulated basement than, say, the garage.

You can vent an electric dryer inside your house during cold months if your home's air is dry and the vent is properly filtered. Its a twofer: dry clothes and warm rooms. (If you have a gas-powered dryer, forget it. It has to be vented to the outside.)

Finally, right now go outside and check the dryer exhaust vent. I mean it. Clean out the vent and make sure the flapper on the outside hood opens and closes easily. If it stays open, you've turned your dryer into an air-conditioner in January. If this isn't what you intended, replace the hood with one that seals very tightly when the dryer's blower is off. Even if the flapper only seals loosely, replace that hood. It's worth the extra bucks to have a good seal.

Washer Energy Savers
Change the way you use your washer and dryer to use less energy and save money.

Use your washer and dryer on the weekends when energy use is lower. If your electric company charges different rates at different times, you'll save money. And the power company won't have to build a new generator for peak loads and raise your rates to pay for it.

Most of the energy used by a washing machine comes from heating the water. So set washer loads for "warm" or "cold" wash instead of hot. Extra dirty loads might need a cold-water pre-soak. The only time I've ever heard a washer really needs hot water is for oily/greasy stains. The washer rinse water should always be cold since the temperature does not affect cleaning. Using cooler water gives you the added bonus of longer-lasting clothes.

Look into cold-water laundry detergents if you have trouble with your current detergents when running cold-water washes.

Run only full, but not overfull, loads in your washer. (If you must be compulsive, weigh a few loads to get a feel for what a "full" load of clothes--per the load capacity listing on your washer--looks like.) One large washer load takes less energy than washing a couple loads on a lower setting. Which reminds me: if you must wash a smaller load, set the water setting lower.

If your washer has a suds-saving feature, use it to save wash water from lightly-soiled loads to reuse in the next load. But only if you're washing the next load right away.

Dishwashers

Whether buying a new dishwasher or using an existing one, you may be able to save a considerable amount of energy by changing the way you operate it.

Here are six tips on how to save energy:

1. Use energy-saving cycles whenever possible.

2. If your dishwasher has a booster heater, turn down your water heater thermostat. Most dishwasher booster heaters can raise water temperature at least 20 F, so a setting of 120 F for your water heater should work fine. The washing cycle will take longer if the dishwasher has to boost the temperature, but unless you need to wash several loads in a row, this shouldn't be a problem.

3. Use the no-heat air-dry feature on your dishwasher if it has one. If you have an older dishwasher that doesn't include this feature, you can turn the dishwasher off after the final rinse cycle is completed an open the door to allow air drying. Using the no-heat dry feature or opening

and air drying the dishes will increase the drying time, and it could lead to increased spotting, according to some in the industry. But try this method sometime to see how well it works with your machine.

4. Don't pre-rinse dishes before putting them in the dishwasher. Modern dishwashers do a superb job of cleaning even heavily soiled dishes. Scrape off food and empty liquids—the dishwasher will do the rest. If you must rinse dishes first, at least use cold water.

5. Wash only full loads. The dishwasher uses the same amount of water whether it's half-full or completely full. Putting dishes in the dishwasher throughout the day and running it once in the evening will use less water and energy than washing the dishes by hand throughout the day.

If you currently wash dishes by hand and fill sinks or plastic tubs with water, it's pretty easy to figure out whether you would use less water with a dishwasher. Simply measure how much water it takes to fill the wash and rinse containers. If you wash dishes by hand two or three times a day, you might be surprised to find out how much water you're currently using. Whether or not you will save energy by switching from washing-by-hand to using a dishwasher depends on both the dishwasher and how you wash the dishes by hand.

6. Load dishes according to manufacturer's instructions. Completely fill the racks to optimize water and energy use, but allow proper water circulation for adequate cleaning.

1.3
SPACE HEATERS AND COOLERS

Portable Heaters

Small space heaters are typically used when the main heating system is inadequate or when central heating is too costly to install or operate. In some cases, small space heaters can be less expensive to use if you only want to heat one room or supplement inadequate heating in one room. They can also boost the temperature of rooms used by individuals who are sensitive to cold, especially elderly persons, without overheating your entire home.

Space heater capacities generally range between 10,000 Btu to 40,000 Btu per hour. Common fuels used for this purpose are: electricity, propane, natural gas, and kerosene.

Although most space heaters rely on convection (the circulation of air in a room) to heat a room, some rely on radiant heating; that is, they emit infrared radiation that directly heats up objects and people that are within their line of sight. Radiant heaters are a more efficient choice when you will be in a room for only a few hours, if you can remain within the line of sight of the heater. They can be more efficient when using a room for a short period because they avoid the energy needed to heat the entire room by instead directly heating the occupant of the room and the occupant's immediate surroundings.

Safety is a top consideration when using space heaters. The U.S. Consumer Product Safety Commission estimates that more than 25,000 residential fires every year are associated with the use of space heaters, causing more than 300 deaths. An estimated 6,000 persons receive hospital emergency room care for burn injuries associated with contacting hot surfaces of room heaters, mostly in non-fire situations.

When buying and installing a small space heater, follow these guidelines:

- Only purchase newer model heaters that have all of the current safety features. Make sure the heater has the Underwriter's Laboratory (UL) label attached to it.
- Choose a thermostatically controlled heaters, since they avoid the energy waste of overheating a room.
- Select a heater of the proper size for the room you wish to heat. Do not purchase oversized heaters. Most heaters come with a general sizing table.
- Locate the heater on a level surface away from foot traffic. Be especially

careful to keep children and pets away from the heater.

Vented and Unvented Combustion Space Heaters

Space heaters are classified as vented and unvented, or "vent free." Unvented combustion units are not recommended for use inside your home, as they introduce unwanted combustion products into the living space, including nitrogen oxides, carbon monoxide, and water vapor. The units also deplete the air in the space where they are located. Most states have banned unvented kerosene heaters for use in the home and at least five have banned the use of unvented natural gas heaters.

Vented units are designed to be permanently located next to an outside wall, so that the flue gas vent can be installed through a ceiling or directly through the wall to the outside. Look for sealed combustion or "100% outdoor air" units, which have a duct to bring outside into the combustion chamber. Sealed combustion heaters are much safer to operate than other types of space heaters, and operate more efficiently because they do not draw in the heated air from the room and exhaust it to the outdoors. They are also less likely to backdraft and adversely affect indoor air quality.

Less expensive (and less efficient) units use the room air for combustion. They do not have a sealed glass front to keep room air away from the fire and should not be confused with a sealed combustion heater.

In addition to the manufacturer's installation and operating instructions, you should follow these general safety guidelines for operating any combustion space heater:

- For liquid-fueled heaters, use only the approved fuel. Never use gasoline! Follow the manufacturer's fueling instructions. Never fill a heater that is still hot. Do not overfill the heater; you must allow for the expansion of the liquid. Only use approved containers clearly marked for that particular fuel, and store them outdoors.
- Have vented space heaters professionally inspected every year. If the heater is not vented properly, not vented at all, or if the vent is blocked, separated, rusted, or corroded, dangerous levels of carbon monoxide can enter the home causing sickness and death. CO also can be produced if the heater is not properly set up and adjusted for the type of gas used and the altitude at which it is installed.

Electric Space Heaters

Electric space heaters are generally more expensive to operate than combustion space heaters, but they are the only unvented space heaters that are safe to operate inside your home. Although electric space heaters avoid indoor air quality concerns, they still carry hazards of potential burns and fires, and should be used with caution.

For convection (non-radiant) space heaters, the best types incorporate a heat transfer liquid, such as oil, that is heated by the electric element. The heat transfer fluid provides some heat storage, allowing the heater to cycle less and to provide a more constant heat source.

When buying and installing an electric space heater, you should follow these general safety guidelines:

- Electric heaters should be plugged directly into the wall outlet. If an extension cord is necessary, use a heavy-duty cord of 14-guage wire or larger.
- For portable electric heaters, buy a unit with a tip-over safety switch, which automatically shuts off the heater if the unit is tipped over.

1.4
LIGHTING AND MAINTENANCE

Lighting Maintenance

Maintenance is vital to lighting efficiency. Light levels decrease over time because of aging lamps and dirt on fixtures, lamps, and room surfaces. Together, these factors can reduce total illumination by 50% or more, while lights continue drawing full power.

The following basic maintenance suggestions can help keep your lights operating at their optimum energy efficiency:

- Clean fixtures, lamps, and lenses every 6-24 months by wiping off the dust. However, never clean an incandescent bulb while it is turned on. The water's cooling effect will shatter the hot bulb.
- Replace lenses if they appear yellow.
- Consider group relamping. Common lamps, especially incandescent and fluorescent lamps, lose 20%-30% of their light output over their service life. Many lighting experts recommend replacing all the lamps in a lighting system at once. This saves labor, keeps illumination high, and avoids stressing any ballasts with dying lamps.

- Clean or repaint small rooms every year and larger rooms every 2-3 years. Dirt collects on surfaces, which reduces the amount of light they reflect.

Lamp and Ballast Replacement for Energy Efficiency

Relamping means substituting one lamp for another to save energy. You can decide to make illumination higher or lower when relamping. Be sure that the new lamp's lumen output fits the tasks performed in the space and conforms to the fixture's specifications.

Matching replacement lamps to existing fixtures and ballasts can be tricky, especially with older fixtures. Buying new fixtures made for new lamps produces superior energy savings, reliability, and longevity compared with relamping.

1.5
LIGHTING CONTROLS

Lighting Controls

Most everyone knows that you can save energy by turning off lights when they're not needed. But sometimes we forget or don't notice that we've left lights on. Lighting controls can be used to automatically turn lights on and off as needed, preventing energy waste.

The most common types of lighting controls include the following:

- Dimmers
- Motion sensors
- Occupancy sensors
- Photosensors
- Timers

Before purchasing and using any lighting controls, it's a good idea to understand basic lighting terms and principles. Also, it helps to explore your lighting options for indoors and/or outdoors if you haven't already. This will help narrow your selection.

Lighting Dimmer Controls

Dimmer controls provide variable indoor lighting for incandescent and fluorescent lamps.

When you dim these lamps, it reduces their wattage and output, which helps save energy.

Off-the-shelf dimmers for incandescent fixtures are inexpensive and provide some energy savings when lights are used at a reduced level.

Dimmers also increase the service life of incandescent lamps significantly. However, dimming incandescent lamps reduces their lumen output more than their wattage. This makes incandescent lamps less efficient as they are dimmed.

Dimming fluorescents requires special dimming ballasts and lamp holders, but does not reduce their efficiency. Fluorescent dimmers are dedicated fixtures and bulbs that provide even greater energy savings than a regular fluorescent lamp.

Lighting Motion Sensor Controls
Motion sensors automatically turn outdoor lights on when they are needed (when motion is detected) and turn them off a short while later. They are very useful for outdoor security and utility lighting provided by incandescent lamps.

Because utility lights and some security lights are needed only when it is dark and people are present, the best way to control might be a combination of motion sensor and photo sensor.

Incandescent flood lights with a photo sensor and motion sensor may actually use less energy than pole-mounted high-intensity discharge (HID) or low-pressure sodium security lights controlled by a photo sensor. Even though HID and low-pressure sodium lights are more efficient than incandescent, they are turned on for a much longer period of time than incandescent using these dual controls.

When turned on, HID and low-pressure sodium lamps can also take up to ten minutes to produce light. Therefore, they don't work well with just a motion sensor.

Lighting Occupancy Sensor Controls
Occupancy sensors—indoor lighting controls—detect activity within a certain area. They provide convenience by turning lights on automatically when someone enters a room.

They reduce lighting energy use by turning lights off soon after the last occupant has left the room.

Occupancy sensors must be located where they will detect occupants or occupant activity in all parts of the room. There are two types of occupancy sensors: ultrasonic and infrared. Ultrasonic sensors detect sound, while infrared sensors detect heat and motion. In addition to controlling ambient lighting in a room, they are useful for task lighting applications, such as over kitchen counters. In such applications, task lights are turned on by the motion of a person washing dishes, for instance, and automatically turn off a few minutes after the person stops.

Lighting Photo sensor Controls
You can use photo sensors to prevent outdoor lights from operating during daylight hours. This can help save energy because you don't have to remember to turn off your outdoor lights.

Photo sensors sense ambient light conditions, making them useful for all types of outdoor lighting. They offer little utility in controlling lights inside the home because lighting needs vary with occupant activity rather than ambient lighting levels.

Lighting Timer Controls
Timers can be used to turn on and off outdoor and indoor lights at specific times.

Simple timers are not often used alone for outdoor lighting because the timer may have to be reset often with the seasonal variation in the length of night.

However, they can be used effectively in combinations with other controls. For example, the best combination for aesthetic (decorative) lighting may be a photo sensor that turns lights on in the evening and a timer that turns the lights off at a certain hour of the night (e.g., 11 P.M.).

For indoor lighting, timers are sometimes used to give unoccupied houses a lived in look.

However, they are an ineffective control for an occupied home because they do not respond to changes in occupant behavior, like occupancy sensors.

1.6
YOUR HOME COMPUTER

Computers

Since July 2007, computers have been rated by the Energy Star ratings program. To be Energy Star-compliant, computers must use a power supply that converts 80 percent of incoming electricity for use by the computer. A basic desktop model must consume less than 50 watts in idle mode. A basic notebook has to consume less than 14 watts of power. A notebook with a graphics chip has to consume less than 22 watts of power.

Did you know that putting your computer in sleep mode when you are not using it can save you approximately $75 per year ~ per computer!

And if you replace your old CRT computer monitor with an LCD display, you can save another $21 per year!

Ask for Energy Star certification

Step1

When shopping, ask for an Energy Star-certified computer or monitor. Using Energy Star Certified computers, monitors, and other devices saves you money on energy costs. It also adds to your efforts to change your lifestyle to a more green way of living. Saving energy helps reduce global warming. As a general rule, laptops are more efficient than desktop models.

Step2

At the Energy Star website, you will find the requirements for an Energy Star Certification. In terms of energy efficiency, the top 25 percent of computers get the Energy Star blessing. Computers are tested for power consumption when in use and when in idle mode.

Step3

The Energy Star computer page provides a link to an Excel document that lists every qualified computer. You can find computers by manufacturer, desktop or laptop system or model. Check this list before you buy, particularly if you are buying online and there is no sales person to help you find an Energy Star certified model.

Step4

After you have your new computer at home, make sure you run it in the most energy efficient mode. Reduce power requirements by closing applications when they aren't in use, by setting the screen brightness at the lowest comfortable level, and by putting it instantly into sleep mode when you know you will be away for a few minutes.

Step5

A computer's energy consumption depends on what you're doing with it. Writing an email uses less energy than playing a graphics-heavy game. I'm not suggesting that you quit playing computer games, but you might think about dragging out the old board games once in a while instead.

Step6

The majority of energy that goes into a computer is actually used during the manufacturing process. Once you have an energy efficient computer, take care to keep it in good running condition for as long as possible. Upgrade and fix when necessary so that the machine runs efficiently for the long term.

Step7

When you get a new energy-efficient computer, dispose of the old one properly. Search in your area for a place that will recycle or reuse the computer parts so that the toxic materials in the old computer don't end up in your local landfill.

When to Turn Off Personal Computers

If you're wondering when you should turn off your personal computer for energy savings, here are some general guidelines to help you make that decision.

Though there is a small surge in energy when a computer starts up, this small amount of energy is still less than the energy used when a computer is running for long periods of time. For energy savings and convenience, consider turning off

- the monitor if you aren't going to use your PC for more than 20 minutes
- both the CPU and monitor if you're not going to use your PC for more than 2 hours.

Make sure your monitors, printers, and other accessories are on a power strip/surge protector. When this equipment is not in use for extended periods, turn off the switch on the power strip to prevent them from drawing power even when shut off. If you don't use a power strip, unplug extra equipment when it's not in use.

Most PCs reach the end of their "useful" life due to advances in technology long before the effects of being switched on and off multiple times have a negative impact on their service life. The less time a PC is on, the longer it will "last." PCs also produce heat, so turning them off reduces building cooling loads.

For cost effectiveness, you also need to consider how much your time is worth. If it takes a long time to shut down the computer and then restart it later, the value of your time will probably be much greater than the value of the amount of electricity you will save by turning off the computer.

Power-Down or Sleep Mode Features

Many PCs available today come with a power-down or *sleep mode* feature for the CPU and monitor. ENERGY STAR® computers power down to a sleep mode that consume 15 Watts or less power, which is around 70% less electricity than a computer without power management features. ENERGY STAR monitors have the capability to power down into two successive "sleep" modes. In the first, the monitor energy consumption is less than or equal to 15 Watts, and in the second, power consumption reduces to 8 Watts, which is less than 10% of its operating power consumption.

Make sure you have the power-down feature set up on your PC through your operating system software. This has to be done by you, otherwise the PC will not power down. If your PC and monitor do not have power-down features, and even if they do, follow the guidelines above about when to turn the CPU and monitor off.

Note: Screen savers are not energy savers. Using a screen saver may in fact use more energy than not using one, and the power-down feature may not work if you have a screen saver activated. In fact, modern LCD color monitors do not need screen savers at all.

1.7
REDUCING YOUR ELECTRICAL CONSUMPTION

A typical household spends about $1,900 a year on energy bills and contributes twice the amount of greenhouse gases to the environment as an average car.

ENERGY STAR, the government-backed symbol for energy efficiency, can guide you in making your home more energy efficient, reducing high energy bills, improving comfort, and protecting the environment—whether you do it yourself or hire a qualified professional.

Common Home Problems and Solutions
Is your home cold, drafty, or uncomfortable? Do you have high energy bills? Ice dams? Peeling paint? Excessive dust? Addressing these types of home problems can make your home more comfortable, and at the same time improve it's energy efficiency — saving you money on utility bills and helping to protect the environment too.

High Energy Bills
High utility bills in summer and winter can often be traced to air leaks in your home's envelope, inefficient windows or heating and

cooling equipment, or poorly sealed and insulated ducts.

Mold, Mildew or Musty Odors
Water leaks or high humidity can lead to mold and mildew. This can cause wood rot, structural damage, peeling paint, and a variety of health problems.

Damp Basement
A damp basement is commonly caused by moisture migrating through the foundation. As this moisture evaporates, it increases indoor humidity and can promote the growth of mold — resulting in an uncomfortable house.

Cold Floors in Winter
Some types of floor coverings (such as wood, stone, tile, or concrete) will naturally feel cold on bare feet. However, insufficient insulation or air infiltration can also cause cold floors.

Drafty Rooms
Cold air coming into or going out of your house, especially through leaks hidden in the attic and basement, can cause rooms to feel drafty and uncomfortable.

Dust
Increased dust could be a sign that it is time to change your air filter or that your ductwork is not well sealed.

Moisture on Windows
Inefficient windows or high indoor moisture levels from air leaks can result in condensation, frost, or pools of water on windows and sills.

Ice Dams
Warm air inside your home leaks into the attic and will warm the underside of the roof causing snow and ice to melt and refreeze as it runs off your roof — forming icicles and ice dams.

Peeling Paint
Peeling or cracking paint on your home's exterior may be a sign of a humidity problem or improper paint application.

Hot or Cold Rooms
Significant differences in temperature from one room to another could be caused by several factors, including inadequate insulation, air leakage, and poor duct performance.

Dry Indoor Air in Winter
Air leaks in your home allow warm humid air to escape and draw in drier colder air.

High Energy Bills

Diagnosis:
One reason for high energy bills is an increase in the price of electricity or heating fuel.

However, it is common to trace high energy bills to an in-efficient component (windows, heating and cooling equipment, ducts insulation) of your home or a failure of one of these components to perform as intended. It is not always easy to pin-point the problem, but fixing it can make your home more energy-efficient and comfortable.

☐ For **best results** hire a contractor who is an energy specialist to do an in-home evaluation. A good specialist will use diagnostic equipment to evaluate the performance of your home and generate a customized list of improvements.

☐ Improvements may include sealing air leaks, adding insulation (Home Sealing) or sealing duct air leaks. Some of these you can do yourself, but you may prefer to hire a contractor.

☐ Turn down the temperature on your water heater to 120 degrees F.

☐ Replace the light bulbs in your highest usage lights with ENERGY STAR CFL bulbs.

☐ When replacing lighting or appliances look for ENERGY STAR qualified light fixtures and appliances

☐ Install an ENERGY STAR qualified programmable thermostat, and use it to save energy while you are away at work.

☐ Contact your utility and ask if they offer any programs to help lower energy bills.

Mold, mildew or musty odors

Diagnosis:
A water leak or high humidity can lead to mold, mildew, or other biological growth. Depending on the severity, conditions can lead to rot, structural damage, premature paint failure, and a variety of health problems. Water can seep into your house from the outside through a leak in your roof, foundation, or small gaps around windows or doors. Water can also come from inside your house from a leaking water pipe, toilet, shower or bathtub. High indoor humidity caused by normal activities of everyday living such as showering, cooking, and drying clothes, can also be a source of mold, mildew or musty odors. Indoor humidity levels between 30% and 50% are ideal. For more information consult EPA's Brief Guide to Mold in your home.

Prescription Checklist:
Where does the problem occur? Attic? Basement? Below a bathroom? Ceiling? Where the problem occurs can lead to what is causing the problem. If the problem is localized (a spot

on the ceiling, wall or corner) it is possibly caused by a water leak. If the problem is in a large area like a whole wall, room or basement then it might be caused by high humidity.

Stop water leaks immediately to minimize the potential mold growth.

- If a leak is the source of your problem, have it fixed first.
- If the leak is in your roof hire a roofing contractor to repair the leak.
- If the leak is from a water pipe, toilet, bathtub or shower, hire a plumber to repair the leak.
- If the leak has caused substantial water damage or mold you will want to hire a contractor who specializes in mold remediation and water damage repairs.
- After repairing the water leak, dry out the area completely.

Reducing indoor humidity
- Do you have a crawlspace under your house? A dirt floor in a crawlspace should be covered with plastic (vapor barrier) to prevent moisture from the soil increasing humidity levels in your home. If there is standing water or the soil is wet, dry it out with fans before covering the floor.

- Use ventilation fans in kitchens and baths to control moisture. Check to make sure ventilation fans venting directly outside. In some cases the vent fan may have been installed to vent into the attic or become disconnected or blocked.
- Your clothes dryer should be vented directly to the outside. Inspect the vent duct. Make sure it is attached securely to the dryer. Check that it is clear of obstructions (e.g. lint). Check for holes that leak air. If vent duct is damaged replace it with a metal duct. The vent duct should be cleaned at least once a year. The Consumer Products Safety Commission additional safety tips for dryer vents.
- Keep air conditioning drip pans clean and the drain lines unobstructed and flowing properly.
- Ask a heating and cooling contractor to check your heating and cooling system to make sure it is sized and operating properly to remove humidity. If you system is too big or the airflow incorrect your air conditioner will not remove humidity like it should. Also, ask the contractor to check your duct system for air leaks, and proper size and air flow to each room.

- Sealing air leaks (Home Sealing) and sealing duct air leaks can help to prevent high humidity levels in your home.

Damp Basement

Diagnosis:
The source of your problem could be a water leak or high humidity. Both can lead to mold, mildew, or other biological growth. Depending on the severity, conditions can lead to rot, structural damage, premature paint failure, and a variety of health problems. Water can seep into your house from the outside through a leak in the foundation, or small gaps around windows or doors. Water can also come from inside your house from a leaking water pipe, toilet, shower or bathtub. High indoor humidity caused by normal activities of everyday living, such as showering, cooking, and drying clothes, can also be a source of your problem. A damp basement is commonly caused by moisture migrating through a concrete foundation. There may not be a sign of any leak or standing water, but the moisture evaporates, increasing indoor humidity. Another common cause is condensation on the cold concrete walls and floors during humid months.

Prescription Checklist:

- Where does the problem occur? Below a bathroom? Ceiling? Corners? Where the problem occurs can lead to what is causing the problem. If the problem is localized (a spot on the ceiling, wall or corner it is possibly caused by a water leak. If the problem is in a large area, like a whole wall or room, then it might be caused by humidity.
- If you suspect a mold problem consult EPA's   Brief Guide to Mold in your home for more information.
- If you plan to remodel your basement, it is important to control moisture problems at the before doing anything else. Corrective actions can be relatively easily but sometimes, depending on the severity of the problem, they can be difficult and expensive.

Stop water leaks

- If a leak is the source of your problem, have it fixed first.
- If you have standing water on the floor of your basement after a heavy rain then it is likely from a leak in the foundation.
  - o Clean rain gutter and redirect downspout runoff away from the foundation.

- o Make sure the ground around the house slopes down away from the foundation. If necessary, re-grade so the ground does slopes away.
- o If you have a sump pump, make sure it is working properly.
- If you have water stains on the ceiling or wall under or near a bathroom it could be a leak from a water pipe, toilet, bathtub or shower. Hire a plumber to repair the leak.
- If the leak has caused substantial water damage or mold you will want to hire a contractor who specializes in mold remediation and water damage repairs.

Reducing indoor humidity

- If your basement has a dirt floor, cover the floor completely with plastic to slow down water vapor coming through the soil.
- Use ventilation fans in kitchens and baths to control moisture. Check to make sure ventilation fans venting directly outside. In some cases the vent fan may have been installed to vent into the attic or become disconnected or blocked.
- Your clothes dryer should be vented directly to the outside. Inspect the vent duct. Make sure it is attached securely to

the dryer. Check that it is clear of obstructions (e.g. lint). Check for holes that leak air. If vent duct is damaged replace it with a metal duct. The vent duct should be cleaned at least once a year. The Consumer Products Safety Commission additional safety tips for dryer vents.

- Ask a heating and cooling contractor to check your heating and cooling system to make sure it is sized and operating properly to remove humidity. If you system is too big or the airflow incorrect your air conditioner will not remove humidity like it should. Also, ask the contractor to check your duct system for air leaks, and proper size and air flow to each room. To help you find a contractor, please refer to our recommendations.
- Sealing air leaks (Home Sealing) and sealing duct air leaks can help to prevent high humidity levels in your home.
- During hot humid months, using a dehumidifier in the basement can reduce condensation on the walls. This may work better after you've sealed air and duct leaks to reduce the amount of humid outdoor air you are bringing into the basement.

Cold Floors

Diagnosis:

Although some types of floor coverings (e.g., wood, stone, tile, or concrete) will naturally feel cold on bare feet, insufficient insulation or air infiltration could be the cause for cold floors.

Common locations:

- Basement floor
- Floor over a garage
- Floor over a crawlspace

Prescription Checklist:

- Air Sealing and insulation (Home Sealing) can help stop drafts and improve the comfort of your home. You can do some things yourself, but for the best solution you need to hire a contractor.
- Contact a heating and cooling contractor to check if your heating and cooling system is providing enough air to each room. Your contractor should check: is a damper closed; has a duct become disconnected from a register; is the duct sized correctly, is the duct leaky?

Drafty Rooms

Diagnosis:

Cold air leaking into your house around windows, doors, electrical outlets, light fixtures, and gaps in corners, can cause rooms to feel drafty and uncomfortable. As cold air is coming in through leaks, warm air is escaping through other leaks. The biggest leaks for escaping air are often found in the attic, and recessed lights are a common location.

Prescription Checklist:

- Air sealing (Home Sealing) can help stop drafts and improve the comfort of your home. The most important leaks are often in the attic. You can do some things yourself, but for the best solution you need to hire a contractor.
- Ask your heating and cooling contractor to check ducts for air leaks and balanced airflow.
- If you have a fireplace, close the damper when not in use.

Dust

Diagnosis:

Dust comes from several sources and is difficult to eliminate completely. Increased dust could be a sign that it is time to change a dirty furnace or air conditioner filter or vacuum cleaner bag. Activities that produce dust (such as sanding) can also be a source an increase. Dust can also be introduced into your home through air leaks in ducts, or air infiltration through leaky doors and windows.

Prescription Checklist:

- Change or clean your furnace and air conditioner filters once a month or according to the filter manufacturer's instructions. Temporarily seal the filter in place with metal-back duct tape. Write the date on the tape with a marker so your know when it was last changed.
- Your clothes dryer should be vented directly to the outside. Inspect the vent duct. Make sure it is attached securely to the dryer. Check that it is clear of obstructions (e.g. lint). Check for holes that leak air. If vent duct is damaged replace it with a metal duct. The vent duct should be cleaned at least once a year. The Consumer Products Safety Commission additional safety tips for dryer vents.
- Consider leaving your shoes at the door so you don't track outside debris-often the largest source of dust -into your house.

- Sealing air leaks (Home Sealing) can help to reduce air infiltration that could be a source of dust.
- Sealing duct air leaks, especially the return duct, can help prevent dust from being circulated throughout your house.
- If you are concerned that dust is coming from your ducts please refer to Should You Have the Air Ducts in Your Home Cleaned? (an EPA publication) for information about duct cleaning.

Moisture on Windows

Diagnosis:
It is difficult to completely eliminate moisture on existing windows. Inefficient windows (e.g., single pane with aluminum frames) or high moisture with inadequate ventilation can result in condensation, frost, or pools of water on windows and sills. Moisture in the air condenses when it touches a cold surface. (The same effect causes a glass of ice tea to "sweat" on a hot humid day.) Continued excess moisture can lead to mold, mildew, and deterioration of your windows and sills.

Prescription Checklist:

To reduce humidity levels in your home:
- Use ventilation fans in kitchens and baths to control moisture.

- Your clothes dryer should be vented directly to the outside. Inspect the vent duct. Make sure it is attached securely to the dryer. Check that it is clear of obstructions (e.g. lint). Check for holes that leak air. If vent duct is damaged replace it with a metal duct. The vent duct should be cleaned at least once a year. The Consumer Products Safety Commission additional safety tips for dryer vents.
- If you have single pane windows, especially with metal frames, install storm windows or consider replacing your existing windows with ENERGY STAR labeled windows.
- If you can't afford to add storm windows or replace your windows now purchase and install a shrink film or polyethylene sheet, window insulation kit from a home center or hardware store.
- If you have a humidifier, check it regularly for proper operation. It could be adding too much moisture to your indoor air.

Ice Dams

Diagnosis:
Ice dams usually occur after a heavy snowfall and several days of freezing temperatures. Warm air inside your home leaks into the attic

and will warm the underside of the roof causing snow and ice on the roof to melt. The melted water will drain along the roof, under the snow, until it reaches the cold overhang. The overhang tends to be at the same temperature as the outdoors and the melted water will refreeze and form an ice dam and icicles. The ice dam can cause damage to the roof, which will result in water leaks to the inside. Frequently the result will be a water spot on the ceiling under the roof damage.

Prescription Checklist:

- Don't get on your roof to solve this problem, it could be dangerous.
- Avoid standing on the ground and "chipping away" at the ice. Not only could this cause damage to your roof, but you can be seriously injured by falling ice, debris, or tools.
- Contacting a roofing contractor to fix your roof leak will not prevent future ice dams.
- Seal air leaks (Home Sealing) and sealing duct air leaks in your attic to stop warm air leakage (the source of the problem).
- After sealing leaks, add additional insulation in your attic.
- Provide adequate attic ventilation so that the underside of the roof and outside air

are at the same temperature. Check to make sure attic insulation is not blocking roof ventilation.

- Clean leaves and other debris from gutters before the first snow. This will help prevent ice build-up in gutters.
- Hire a contractor who is an energy specialist or specializes in air sealing to do an in-home evaluation. A good specialist will use diagnostic equipment to evaluate the performance of your home and generate a customized list of improvements.

Peeling Paint

Diagnosis:
Peeling or cracking paint, on your home's exterior, may be a sign of a humidity problem or improper application. Peeling exterior paint is caused by moisture being absorbed through the back of wood siding and passing through to the exterior surface under the paint. The paint loses adhesion and peels off. The exterior should be vented to allow any moisture behind the siding to escape.

Prescription Checklist:

- Control moisture problems.
- Air sealing (Home Sealing) can keep moist air from leaking through your

walls. To adequately prevent moist air from moving into wall cavities, you may need to hire a contractor who is a building science specialist.

- Apply primer to surfaces before painting and follow the paint manufacturer's application instructions.

Hot or Cold Room

Diagnosis:
Temperature differences of up to three degrees from room to room are not uncommon, but often one or several rooms are uncomfortably warm or cold. This condition could be caused by several factors within your home including inadequate insulation, air leakage, poor duct system design, duct leakage, unwanted heating by the sun in warmer months, or a failure in part of your heating and cooling system.

Common problem rooms include:

- Attic
- Room over a garage
- Basement
- Additions

Prescription Checklist:
For best results hire a contractor who is an energy specialist to do an in-home evaluation. A good specialist will use diagnostic equipment to

evaluate the performance of your home and generate a customized list of improvements.

- Ask your contractor to check if your heating and cooling system is operating correctly.
- Ask your contractor to check your ducts for air leakage and proper distribution of air.
- Seal air leaks and add insulation (Home Sealing).
- If the sun is making rooms to hot, consider shades or solar screening.
- After trying these items, consider ENERGY STAR labeled ceiling fans to make room air circulation more uniform. You will need to hire an electrician to install it.

Dry Air

Diagnosis:
Air leaks in your home allow warm humid air to escape and draw in drier colder air. Dry indoor air can contribute to dry throat and skin and static shocks. Proper humidity levels keep furniture and your home from drying out and reduce the energy use of your heating system because you will feel warmer at a lower thermostat setting.

Prescription Checklist:

- Ask a heating and cooling contractor to check your heating and cooling system to make sure it is operating properly. Also, ask the contractor to check your duct system for air leaks, and proper size and air flow to each room.
- Seal air leaks (Home Sealing) to prevent infiltration of cold, dry air for outside. If you have a tight home you may not need a humidifier.

PART 2

NATURAL GAS / PROPANE

2.1

INSULATION IN YOUR HOME

Insulation & Air Sealing

You can reduce your home's heating and cooling costs by as much as 30 percent through proper insulation and air sealing techniques. These techniques will also make your home more comfortable. Reducing your home heating and cooling bills begins with conducting a home energy audit to assess where your home may be losing energy through air leaks or inadequate insulation.

- Remember that new windows must be installed correctly to avoid air leaks around the frame. Look for a reputable, qualified installer.
- In temperate climates with both heating and cooling seasons, select windows with both low U-values and low solar heat gain coefficiency (SHGC) to maximize energy benefits.
- In temperate climates with both heating and cooling seasons, select windows with both low U-values and low solar heat gain coefficiency (SHGC) to maximize energy benefits.

- Select windows with air leakage ratings of 0.3 cubic feet per minute or less.
- Remember, the lower the U-value, the better the insulation. In colder climates, a U-value of 0.35 or below is recommended. These windows have at least double glazing and a low-e coating.
- When you're shopping for new windows, look for the National Fenestration Rating Council label; it means the window's performance is certified.
- Installing new, high-performance windows will improve your home's energy performance. While it may take many years for new windows to pay off in energy savings, the benefits of added comfort and improved aesthetics and functionality may make the investment worth it to you.
- Apply sun-control or other reflective films on south-facing windows to reduce solar gain.
- Install awnings on south- and west-facing windows.
- Close curtains on south- and west-facing windows during the day.
- Install white window shades, drapes, or blinds to reflect heat away from the house.
- Repair and weatherize your current storm windows, if necessary.

- Install exterior or interior storm windows; storm windows can reduce heat loss through the windows by 25% to 50%. Storm windows should have weatherstripping at all moveable joints; be made of strong, durable materials; and have interlocking or overlapping joints. Low-e storm windows save even more energy.
- Keep windows on the south side of your house clean to let in the winter sun.
- Close your curtains and shades at night; open them during the day.
- Install tight-fitting, insulating window shades on windows that feel drafty after weatherizing.
- You can use a heavy-duty, clear plastic sheet on a frame or tape clear plastic film to the inside of your window frames during the cold winter months. Remember, the plastic must be sealed tightly to the frame to help reduce infiltration.
- Conduct an energy audit of your home to find air leaks and to check for the proper level of insulation. Common sources of air leaks include cracks around windows and doors, gaps along baseboard, mail chutes, cracks in brick, siding, stucco or foundation, or where any external lines (phone, cable, electric, and gas) enter the home.

- To test for air leaks on your own, on a windy day, hold a lit candle next to windows, doors, electrical outlets, or light fixtures to test for leaks. Also, tape clear plastic sheeting to the inside of your window frames if drafts, water condensation, or frost are present.
- Plug air leaks with caulking, sealing, or weather stripping to save 10 percent or more on your energy bill.
- Adequate insulation in your attic, ceilings, exterior and basement walls, floors, and crawlspaces, as recommended for your geographical area, can save you up to 30 percent on home energy bills.
- Installing storm windows over single-pane windows or replacing them with ENERGY STAR® windows can reduce heat loss from air leakage, and reflect heat back into the room during the winter months to save even more energy.
- In cold climates, ENERGY STAR® windows can reduce your heating bills by 30 to 40 percent compared to uncoated, single-pane windows, according to the Efficient Windows Collaborative.
- Close fireplace dampers when not in use. A chimney is designed for smoke to escape, so until you close it, warm air escapes.

2.2
WINTER STORM TIPS

Winter's arrival brings with it the possibility of severe weather.

Consumers are warned to keep safety as their top priority and to always stay away from any downed power lines that may result from winter storms.

When a power outage occurs, consumers are encouraged to first check breaker panels or fuse boxes for tripped circuit breakers or blown fuses. If the home's electrical system is intact, determine the extent of the outage by checking to see whether neighbors also have a problem.

Don't assume the utility knows you are without power. Your report and those of your neighbors help the utility identify the scope of power outages and aid in electric service restoration efforts.

Following are some suggestions for preparing for and coping with a winter power outage:

Remove dead or dying ash trees

Dead and dying ash trees are a widespread problem due to the emerald ash borer, an insect that has invaded ash trees in several states. As a result of this infestation, ash trees, which may grow to heights of 60 feet, have become structurally unstable and may fall at any time, especially during a storm.

People should remove these trees from their property. Although there is an expense involved with the tree removal, the potential for injury or death and damage to homes, vehicles or property is far worse.

Steps to take before a storm

-- Assemble an emergency kit that is easily accessible. It should include a battery-powered radio or television, a flashlight and extra batteries, candles and matches or a lighter, a first-aid kit, a fire extinguisher, bottled water and non-perishable food. In addition, keep a corded or cell phone on hand because cordless telephones need electricity to operate.

-- Select a small, well-insulated room with a fireplace, wood stove or fuel-burning heater to use as emergency living quarters. Keep an emergency supply of fuel or wood handy. For safety, always store fuel in a dry place away from the house. In case of extended outages, blankets or cardboard can be hung over windows and doors to minimize heat loss.

After a storm

-- Stay at least 20 feet away from a downed power line and anything it contacts, especially metal fences. Treat every downed power line as if it is energized and keep children and pets out of the area.

-- Don't open refrigerators and freezers more often than absolutely necessary. A closed refrigerator will stay cold for 12 hours. Kept closed, a well-filled freezer will preserve food for two days. Partially thawed food or food that has ice crystals usually can be refrozen. Open faucets slightly so they constantly drip to prevent pipes from freezing.

-- A fuel-burning heater -- such as kerosene -- requires an area with proper ventilation to prevent buildup of harmful fumes. Keep portable heaters away from furniture, draperies and other flammable materials.

-- Never use a propane or charcoal grill as an indoor heating or cooking source.

-- Turn off or unplug all major appliances to prevent an electrical overload when power is restored. Leave on one light to indicate when power is restored.

-- If using a portable generator pull or switch to "off" all main fuses or circuit breakers to protect line crews working to restore service. Always operate generators outdoors to avoid dangerous buildup of toxic fumes.

-- During low-voltage conditions -- when lights are dim and television pictures are smaller -- shut off motor-driven appliances such as refrigerators to prevent overheating and possible damage. Sensitive electronic devices also should be unplugged.

-- When clearing snow and ice from roofs and gutters, be sure to inspect the area for overhead power lines. Maintain a 20-foot clearance between the power lines and your ladder and tools. Contact with overhead lines can be deadly.

2.3
REDUCING NATURAL GAS AND PROPANE CONSUMPTION

Get a head start on cutting your heating bill

- Seal up. Save on your heating bill by reducing the air leaks in your home – caulk, seal and weather-strip all seams, cracks and openings to the outside. Also, replace cracked and peeling caulk around windows, doors and siding.

- Insulate. Improving wall, ceiling and basement/crawl space insulation can significantly reduce heating costs

- Get a check up. A well maintained heating system runs more efficiently. Have your furnace inspected annually by a licensed and reputable heating contractor, ideally before the heating season begins.

- Clean or replace filters regularly. A clogged or dirty filter can restrict proper airflow and force your furnace to work harder.

PART 3

YOUR WATER HEATER

3.1
EFFICIENT WATER HEATING

Energy-Efficient Water Heating

To lower your water heating bills, try one or more of these energy-saving strategies:

- Reduce your hot water use
- Lower your water heating temperature
- Insulate your water heater tank
- Insulate hot water pipes
- Install heat traps on a water heater tank
- Install a timer and use off-peak power for an electric water heater
- Install a drain-water heat recovery system.

3.2
DRAIN RECOVERY SYSTEMS

Drain-Water Heat Recovery
Any hot water that goes down the drain carries away energy with it. That's typically 80%-90% of the energy used to heat water in a home. Drain-water (or grey water) heat recovery systems capture this energy to preheat cold water entering the water heater or going to other water fixtures.

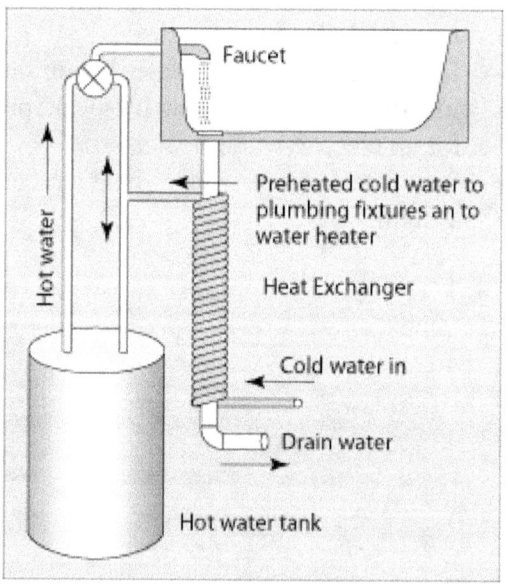

Faucet

Preheated cold water to plumbing fixtures an to water heater

Heat Exchanger

Hot water

Cold water in

Drain water

Hot water tank

How It Works
Drain-water heat recovery technology works well with all types of water heaters, especially

with demand and solar water heaters. Also, drain-water heat exchangers can recover heat from the hot water used in showers, bathtubs, sinks, dishwashers, and clothes washers. They generally have the ability to store recovered heat for later use. You'll need a unit with storage capacity for use with a dishwasher or clothes washer. Without storage capacity, you'll only have useful energy during the simultaneous flow of cold water and heated drain water, like while showering.

Some storage-type systems have tanks containing a reservoir of clean water. Drain water flows through a spiral tube at the bottom of the heat storage tank. This warms the tank water, which rises to the top. Water heater intake water is preheated by circulation through a coil at the top of the tank.

Non-storage systems usually have a copper heat exchanger that replaces a vertical section of a main waste drain. As warm water flows down the waste drain, incoming cold water flows through a spiral copper tube wrapped tightly around the copper section of the waste drain. This preheats the incoming cold water that goes to the water heater or a fixture, such as a shower.

By preheating cold water, drain-water heat recovery systems help increase water heating capacity. This increased capacity really helps if you have an undersized water heater. You can also lower your water heating temperature without affecting the capacity.

Cost and Installation

Prices for drain-water heat recovery systems range from $300 to $500. You'll need a qualified plumbing and heating contractor to install the system. Installation will usually be less expensive in new home construction. Paybacks range from 2.5 to 7 years, depending on how often the system is used.

3.3
HEAT TRAPS FOR YOUR TANK

Install Heat Traps on a Water Heater Tank for Energy Savings

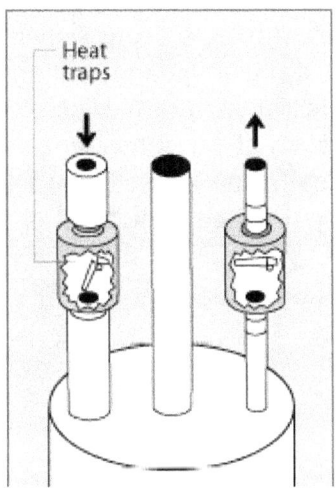

If your storage water heater doesn't have heat traps, you can save energy by adding them to your water heating system. They can save you around $15-$30 on your water heating bill by preventing convective heat losses through the inlet and outlet pipes.

Heat trap valves or loops of pipe allow water to flow into the water heater tank but prevent unwanted hot-water flow out of the tank. The valves have balls inside that either float or sink into a seat, which stops convection. These specially designed valves come in pairs. The valves are designed differently for use in either the hot or cold water line.

A pair of heat traps costs only around $30. However, unless you can properly solder a pipe joint, heat traps require professional installation by a qualified plumbing and heating contractor. Therefore, heat traps are most cost effective if they're installed at the same time as the water heater. Today, many new storage water heaters have factory-installed heat traps or have them available

3.4
TIMERS AND PEAK TIMES

Install a Timer and Use Off-Peak Power for Electric Water Heaters

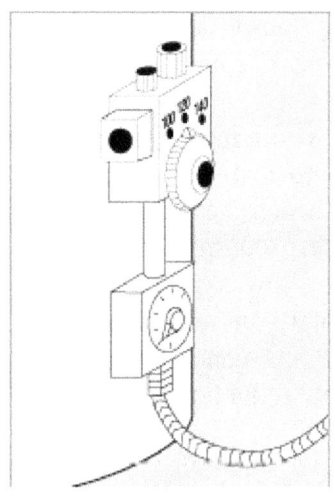

If you have an electric water heater, you can save an additional 5%-12% of energy by installing a timer that turns it off at night when you don't use hot water and/or during your utility's peak demand times.

You can install a timer yourself. They can cost $60 or more, but they can pay for themselves in about 1 year. Timers are most cost effective if you don't want to install a heat trap and insulate your water heater tank and pipes. Timers aren't as cost effective or useful on gas water heaters because of their pilot lights.

Contact your utility to see if it offers a demand management program. Some utilities offer "time of use" electricity rates that vary according to the demand on their system. They charge higher rates during "on-peak"< times and lower rates during "off-peak" times.

Some even offer incentives to customers who allow them to install control devices that shut off electric water heaters during peak demand periods. These control devices may use radio signals that allow a utility to shut off a water heater remotely anytime demand is high. Shut-off periods are generally brief so customers experience no reduction in service.

3.5
INSULATING YOUR TANK

Insulate Your Water Heater Tank for Energy Savings

Unless your water heater's storage tank already has a high R-value of insulation (at least R-24), adding insulation to it can reduce standby heat losses by 25%-45%. This will save you around 4%-9% in water heating costs.

If you don't know your water heater tank's R-value, touch it. A tank that's warm to the touch needs additional insulation.

Insulating your storage water heater tank is fairly simple and inexpensive, and it will pay for itself in about a year. You can find pre-cut jackets or blankets available from around $10-$20. Choose one with an insulating value of at least R-8. Some utilities sell them at low prices, offer rebates, and even install them at a low or no cost.

Insulating an Electric Water Heater Tank

You can probably install an insulating pre-cut jacket or blanket on your electric water heater tank yourself. Read and follow the directions carefully. Leave the thermostat access panel(s) uncovered. Don't set the thermostat above 130°F

on electric water heater with an insulating jacket or blanket
the wiring may overheat.

You may want to see our instructions for installing an insulation blanket on an electric water heater.

You also might consider placing a piece of rigid insulation bottom board under the tank of your electric water heater. This will help prevent heat loss into the floor, saving another 4%-9% of water heating energy.

It's best done when installing a new water heater.

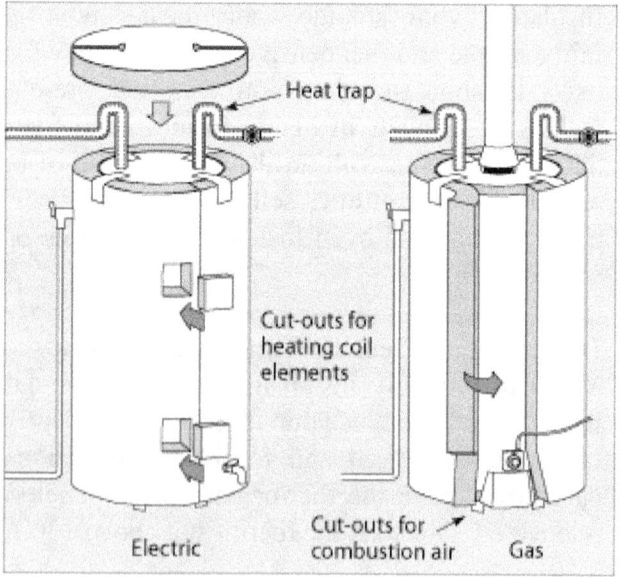

Insulating a Gas Water Heater Tank
The installation of insulating blankets or jackets on gas and oil-fired water heater tanks is more difficult than those for electric water heater tanks. It's best to have a qualified plumbing and heating contractor add the insulation. If you want to install it yourself, read and follow the directions very carefully. Keep the jacket or blanket away from the drain at the bottom and the flue at the top. Make sure the airflow to the burner isn't obstructed. Leave the thermostat uncovered, and don't insulate the top of a gas water heater tank the insulation is combustible and can interfere with the draft diverter.

Insulate Hot Water Pipes for Energy Savings
Insulating your hot water pipes reduces heat loss and can raise water temperature 2°-4°F hotter than uninsulated pipes can deliver, allowing for a lower water temperature setting. You also won't have to wait as long for hot water when you turn on a faucet or showerhead, which helps conserve water.

Insulate all accessible hot water pipes, especially within 3 feet of the water heater. It's also a good idea to insulate the cold water inlet pipes for the first 3 feet.

Use quality pipe insulation wrap, or neatly tape strips of fiberglass insulation around the pipes.

Pipe sleeves made with polyethylene or neoprene foam are the most commonly used insulation. Match the pipe sleeve's inside diameter to the pipe's outside diameter for a snug fit. Place the pipe sleeve so the seam will be face down on the pipe. Tape, wire, or clamp (with a cable tie) it every foot or two to secure it to the pipe. If you use tape, some recommend using acrylic tape instead of duct tape.

On gas water heaters, keep insulation at least 6 inches from the flue. If pipes are within 8 inches of the flue, your safest choice is to use fiberglass pipe-wrap (at least 1-inch thick) without a facing. You can use either wire or aluminum foil tape to secure it to the pipe.

3.6

LOWERING YOUR TANKS ENERGY CONSUMPTION

Water Heating

Water heating can account for 14%-25% of the energy consumed in your home. You can reduce your monthly water heating bills by selecting the appropriate water heater for your home or pool and by using some energy-efficient water heating strategies.

Lower Water Heating Temperature for Energy Savings

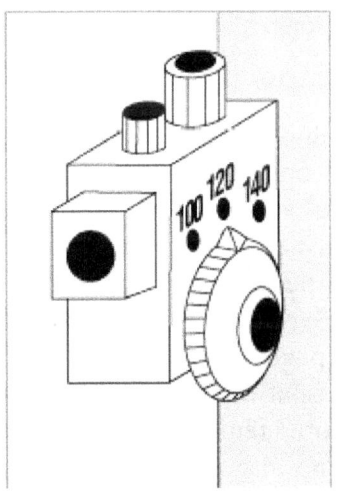

You can reduce your water heating costs by simply lowering the thermostat setting on your water heater. For each 10°F reduction in water temperature, you can save between 3%-5% in energy costs.

Although some manufacturers set water heater thermostats at 140°F, most households usually only require them set at 120°F. Water heated at 140°F also poses a safety hazard scalding. However, if you have a dishwasher without a booster heater, it may require a water temperature within a range of 130°F to 140°F for optimum cleaning.

Reducing your water temperature to 120°F also slows mineral buildup and corrosion in your water heater and pipes. This helps your water heater last longer and operate at its maximum efficiency.

Consult your water heater owner's manual for instructions on how to operate the thermostat. You can find a thermostat dial for a gas storage water heater near the bottom of the tank on the gas valve. Electric water heaters, on the other hand, may have thermostats positioned behind screw-on plates or panels. As a safety precaution, shut off the electricity to the water heater before removing/opening the panels. Keep in mind that an electric water heater may have two thermostats one each for the upper and lower heating elements.

Mark the beginning temperature and the adjusted temperature on the thermostat dial for future reference.

After turning it down, check the water temperature with a thermometer at the tap farthest from the water heater. Thermostat dials are often inaccurate. Several adjustments may be necessary before you get the right temperature.

If you plan to be away from home for at least 3 days, turn the thermostat down to the lowest setting or completely turn off the water heater. To turn off an electric water heater, switch off the circuit breaker to it.

For a gas water heater, make sure you know how to safely relight the pilot light before turning it off.

3.7
REDUCING HOT WATER USE

Reduce Hot Water Use for Energy Savings
You can lower your water heating costs by using and wasting less hot water in your home. To conserve hot water, you can fix leaks, install low-flow fixtures, and purchase an energy-efficient dishwasher and clothes washer.

Fix Leaks
You can significantly reduce hot water use by simply repairing leaks in fixtures faucets and showerheads or pipes. A leak of one drip per second can cost $1 per month.

If your water heater's tank leaks, you need a new water heater.

Install Low-Flow Fixtures
Federal regulations mandate that new showerhead flow rates can't exceed more than 2.5 gallons per minute (gpm) at a water pressure of 80 pounds per square inch (psi). New faucet flow rates can't exceed 2.5 gpm at 80 psi or 2.2 gpm at 60 psi. You can purchase some quality, low-flow fixtures for around $10 to $20 a piece and achieve water savings of 25-60%.

Showerheads

For maximum water efficiency, select a shower head with a flow rate of less than 2.5 gpm.

There are two basic types of low-flow showerheads: aerating and laminar-flow. Aerating showerheads mix air with water, forming a misty spray. Laminar-flow showerheads form individual streams of water.

If you live in a humid climate, you might want to use a laminar-flow showerhead because it won't create as much steam and moisture as an aerating one.

Before 1992, some showerheads had flow rates of 5.5 gpm. Therefore, if you have fixtures that pre-date 1992, you might want to replace them if you're not sure of their flow rates. Here's a quick test to determine whether you should replace a showerhead:

1. Place a bucket marked in gallon increments under your shower head.
2. Turn on the shower at the normal water pressure you use.
3. Time how many seconds it takes to fill the bucket to the 1-gallon (3.8 liter) mark.

If it takes less than 20 seconds to reach the 1-gallon mark, you could benefit from a low-flow shower head.

Faucets

The aerator the screw-on tip of the faucet ultimately determines the maximum flow rate of a faucet. Typically, new kitchen faucets come equipped with aerators that restrict flow rates to 2.2 gpm, while new bathroom faucets have ones that restrict flow rates from 1.5 to 0.5 gpm.

Aerators are inexpensive to replace and they can be one of the most cost-effective water conservation measures. For maximum water efficiency, purchase aerators that have flow rates of no more than 1.0 gpm. Some aerators even come with shut-off valves that allow you to stop the flow of water without affecting the temperature. When replacing an aerator, bring the one you're replacing to the store with you to ensure a proper fit.

Purchase Energy-Efficient Dishwashers and Clothes Washers

The biggest cost of washing dishes and clothes comes from the energy required to heat the water. You'll significantly reduce your energy costs if you purchase and use an energy-efficient dishwasher and clothes washer.

Dishwashers

It's commonly assumed that washing dishes by hand saves hot water. However, washing dishes by hand several time a day can be more expensive than operating an energy-efficient dishwasher. You can consume less energy with an energy-efficient dishwasher when properly used and when only operating it with full loads.

When purchasing a new dishwasher, check the EnergyGuide label to see how much energy it uses. Dishwashers fall into one of two categories: compact capacity and standard capacity. Although compact-capacity dishwashers may appear to be more energy efficient on the EnergyGuide Label, they hold fewer dishes, which may force you to use it more frequently. In this case, your energy costs could be higher than with a standard-capacity dishwasher.

One feature that makes a dishwasher more energy efficient is a booster heater. A booster heater increases the temperature of the water entering the dishwasher to the 140°F recommended for cleaning. Some dishwashers have built-in boosters, while others require manual selection before the wash cycle begins.

Some also only activate the booster during the heavy-duty cycle. Dishwashers with booster heaters typically cost more, but they pay for themselves with energy savings in about 1 year if you also lower the water temperature on your water heater.

Another dishwasher feature that reduces hot water use is the availability of cycle selections. Shorter cycles require less water, thereby reducing energy cost.

If you want to ensure that your new dishwasher is energy efficient, purchase one with an ENERGY STAR® label.

Clothes Washers
Unlike dishwashers, clothes washers don't require a minimum temperature for optimum cleaning.

Therefore, to reduce energy costs, you can use either cold or warm water for most laundry loads. Cold water is always sufficient for rinsing.

Inefficient clothes washers can cost three times as much to operate than energy-efficient ones. Select a new machine that allows you to adjust the water temperature and levels for different loads. Efficient clothes washers spin-dry your clothes more effectively too, saving energy when drying as well. Also, front-loading machines use less water and, consequently, less energy than top loaders.

Small-capacity clothes washers often have better EnergyGuide label ratings. However, a reduced capacity might increase the number of loads you need to run, which could increase your energy costs.

If you want to ensure that your new clothes washer is energy efficient, purchase one with an ENERGY STAR label.

Water Heating

Water heating can account for 14%–25% of the energy consumed in your home. You can reduce your monthly water heating bills by selecting the appropriate water heater for your home or pool and by using some energy-efficient water heating strategies.

- Wash only full loads of dishes and clothes.
- Take short showers instead of baths.
- Lower the thermostat on your hot water heater to 120° F.
- You might qualify for tax credits or rebates for buying a solar water heater. Visit the Database of State Incentives for Renewable Energy Web site and see.
- Heat pump water heaters are very economical in some areas.
- Consider natural-gas on-demand or tankless water heaters. Researchers have

found savings can be up to 30% compared with a standard natural-gas storage tank water heater.

- Consider installing a drain water waste heat recovery system. A recent DOE study showed energy savings of 25% to about 30% for water heating using such a system.

- Buy a new energy-efficient water heater. While it may cost more initially than a standard water heater, the energy savings will continue during the lifetime of the appliance. Look for the EnergyGuide label.

- Although most water heaters last 10-15 years, it's best to start shopping for a new one if yours is more than 7 years old. Doing some research before your heater fails will enable you to select one that most appropriately meets your needs.

- Drain a quart of water from your water tank every 3 months to remove sediment that impedes heat transfer and lowers the efficiency of your heater. The type of water tank you have determines the steps to take, so follow the manufacturer's advice.

- Install heat traps on the hot and cold pipes at the water heater to prevent heat loss. Some new water heaters have built-in heat traps.

- If you are in the market for a new dishwasher or clothes washer, consider buying an efficient, water-saving ENERGY STAR® model to reduce hot water use.
- Insulate the first 6 feet of the hot and cold water pipes connected to the water heater.
- Insulate your natural gas or oil hot-water storage tank, but be careful not to cover the water heater's top, bottom, thermostat, or burner compartment. Follow the manufacturer's recommendations; when in doubt, get professional help.
- Insulate your electric hot-water storage tank, but be careful not to cover the thermostat. Follow the manufacturer's recommendations.
- Take more showers than baths. Bathing uses the most hot water in the average household.
- Lower the thermostat on your water heater; water heaters sometimes come from the factory with high temperature settings, but a setting of 120°F provides comfortable hot water for most uses.
- Repair leaky faucets promptly; a leaky faucet wastes gallons of water in a short period of time.
- Install aerating, low-flow faucets and showerheads.

- Select a shower head with a flow rate of less than 2.5 gpm (gallons per minute) for maximum water efficiency. Before 1992, some showerheads had flow rates of 5.5 gpm, so you might want to replace them if you're not sure of their flow rates.
- Insulate your hot water pipes, which will reduce heat loss and can raise water temperature 2°F–4°F hotter than uninsulated pipes. This allows for a lower water temperature setting.
- Lowering the thermostat on your water heater by 10°F can save you between 3%–5% in energy costs. Most households only require a water heater thermostat setting of 120°F, or even 115°F.
- Did you know that 85-90% of the energy from hot water is wasted when it goes down the drain? Install a drain-water heat recovery system to pre-heat new water using the heat from drained water.
- If heating a swimming pool, consider a swimming pool cover. Evaporation is by far the largest source of energy loss in swimming pools.

PART 4

HUNDREDS OF ENERGY SAVING TIPS

4.1

HOW MUCH ENERGY AM I USING?

How do I find out how much electricity something uses?

The shortcut is to just look at the label!

Nearly everything you can plug into the wall has a label that says how much electricity it uses. (It may be printed directly into the plastic or metal.) You may have to hunt for the label. It's often located on the bottom or side of the device, or possibly where the power cord enters the unit. If the device is powered with an AC/DC adapter, the electrical rating is usually listed on the adapter itself.

If the label only gives the number of amps and not the number of watts, then just multiply the amps by 120 to get the number of watts. (Amps x Volts = Watts, and most U.S. electricity is 120 volts. So a hot plate that uses 6 amps uses 6 x 120 = 720 watts. Most other countries use 240 volts instead of 120, so outside of North America and Japan use 240 instead of 120 in your calculations.) Note that if a device is powered by a transformer (one of those great big plugs), then the transformer has converted the electricity from AC to DC, so you need to multiply by the DC voltage, not the AC voltage of 120. For example, if the device says "INPUT 9V, 0.5A", then that's 9 volts x 0.5 amps = 4.5 watts.

You may have noticed that appliances may be labeled 110, 115, or 120 volts. Appliances are actually designed to accept a range of voltages, between 110-120 volts, and the exact voltage coming out of your electrical socket can vary depending on conditions at the power plant and in your own home. Let's just agree that when we say 120 volts, we understand that it's actually a range from 110-120. And just use 120 for your calculations (unless you're outside of North America or Japan, in which case you probably have 240 volts).

Your device might actually list a huge voltage range, like 100-240V. That just means that it will work with any country's voltage. For your calculations, use the voltage for the country where you're plugging the device in.

Some important caveats:

1. **The amount of electricity listed on the label is the *maximum* amount that the appliance will ever use.** For example, a 300-watt refrigerator will only run at 300 watts when the compressor's running (which is when it makes that humming

sound, indicating that it's actually chilling the air inside). Most of the time the fridge just sits there, using only 5 watts or so for its electronics. If the amount of work done by a device varies up and down, then so does its energy use. (e.g., a stereo that can be turned up or down, an oven that can be set at various temperatures, a fridge that sometimes runs and sometimes doesn't, a computer that sometimes spins its various drives and sometimes has to use more of its brainpower, etc.) The label on computers is particularly useless; a computer labeled at 300 watts probably uses only about 65.

2. **Many consumer items are advertised according to their power *output*, not *input*.** That means the stereo that says 30 watts on the box might actually require 50 watts to make 30 watts of sound (assuming the volume was cranked), and your 900-watt microwave oven might actually use 1400 watts (on its highest setting). That's because all electrical devices are inefficient -- they have to use some extra energy to do what they do.

• Energy Hogs

4400 watts	Clothes dryer (electric)
4400 watts	Electric oven
3800 watts	Water heater (electric)
3500 watts	Central Air Conditioner (2.5 tons)
1500 watts	Microwave oven
1500 watts	Toaster (four-slot)
900 watts	Coffee maker
800 watts	Range burner
500-1440 watts	Window unit air conditioner
200-700 watts	Refrigerator
60-100 watts	Light bulb (energy hog because houses have lots of lights, and it's easy to leave them on when they're not being used)

Fans

100 watts	Floor fan or box fan (high speed)
15-95 watts	Ceiling fan (Bigger fans and faster speeds use more energy. My 2004 42" Hampton Bay uses 24/28/42 watts on low/med/high respectively, according to the manual. Progress Energy says on high speed fans use 55/75/95 watts for 36"/48"/52" models respectively.)

Computers

140-330 watts	Desktop Computer & 17" CRT monitor
1-20 watts	Desktop Computer & Monitor

	(in sleep mode)
120 watts	17" CRT monitor
40 watts	17" LCD monitor
45 watts	Laptop computer

Other

60-100 watts	Regular light bulb
4-165 watts	Video game (While playing game, 30W for PS2, 70W for XBox, and 165W for XBox 360.
55-90 watts	19" television
18 watts	Compact fluorescent light bulb
4 watts	Clock radio
3 watt-hours	Total power stored by an alkaline AA battery. This is to put batteries into perspective. If you could power your clock radio with a AA battery, it wouldn't even last an hour.

3. **Knoknowing how much electricity a device uses at a given moment doesn't tell you how much it's using in a month, because it's probably not running 24/7** (and if it *is* running 24/7 like a fridge, it's probably not using the maximum amount of electricity, as we discussed earlier).

4. **Some devices use a small amount of electricity even when they're not on.** For example, VCR's and microwaves draw a small amount to power the time display. This amount is often 5 watts or less. Devices which run off transformers also draw a small amount of power.

And of course, electricity consumption of a device varies from brand to brand, and condition to condition.

Electrical usage of household items

Naturally, electrical usage will vary from model to model, so remember that the table shown are just *examples*.

But now that you know how to find out how much electricity things use (from the previous section), it's best to make your own table. Remember that the maximum amount of electricity your appliance uses will be printed on the appliance!

U.S. household use of electricity, 2001

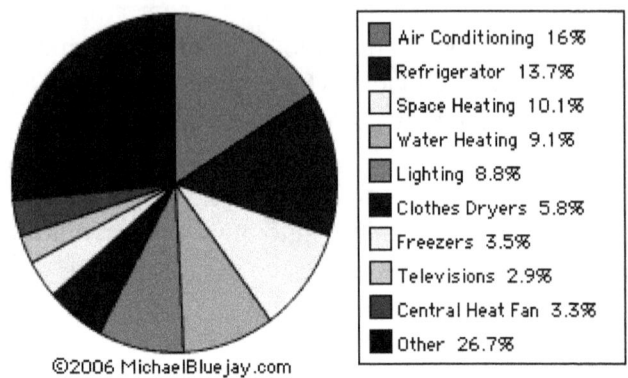

■	Air Conditioning 16%
■	Refrigerator 13.7%
☐	Space Heating 10.1%
■	Water Heating 9.1%
■	Lighting 8.8%
■	Clothes Dryers 5.8%
☐	Freezers 3.5%
■	Televisions 2.9%
■	Central Heat Fan 3.3%
■	Other 26.7%

©2006 MichaelBluejay.com

Estimating use per month

Of course, knowing that your refrigerator uses, say, 500 watts when the compressor's on doesn't tell you how much energy it uses in a month, because the compressor's not on 24/7. Here are some websites that give sample costs for various household items considering how much those items are used:

Which devices use how much?

The chart above shows how the average home used electricity in 2001. (Source: Dept. of Energy; see also Energy Info. Administration & Energy.gov)

Of course, air conditioning uses a bigger chunk of the pie in the summer. According to Austin Energy, AC accounts for 60-70% of the average home's summertime power bill.

Here's how much various strategies can save you.

Easy Strategies

Strategy	Up front cost	Savings per year
(1) Use space heaters to heat only the rooms you're in, rather than a central system that heats the whole house, and turning off the heat when you're not home.	$80	**$1286**
(2) Use ceiling fans instead of the air conditioner	$100 if you don't already have ceiling fans	**$665**
(3) Wash laundry in cold water instead of hot or warm	none	**$167**
(4) Use a clothesline or a laundry rack instead of a	$20	**$141**

dryer

(5) Turn off a single 100-watt light bulb, from running constantly	$0	**$96**
(6) Replace regular light bulbs with compact fluorescents	$32	**$90**
(7) Sleep your computer when you're not using it	$0	**$73**
Total	**$232** *once*	**$2518** *every year*

Aggressive Strategies

(8) Replace top-loading washer with front-loading washer	$500	$90
(9) Replace 1990 fridge with 2004 model	$300	**$45**
(10) Replace a CRT computer monitor in a home office with an LCD display	$200	**$21**
Total	**$1000** *once*	**$156** *every year*

Assumptions:
National average electrical rate of 11¢/kWh.
(1) One 5000-watt central system running 24/7 for four months, vs. two 1500-watt heaters running 8 hours a day

for   four   months.
(2) Stop running a 3500-watt AC 12 hours a day for five months, use two large ceiling fans instead, 12 hours/day.
(3) 1/3 of loads originally on the Hot/Warm setting and 2/3 on Warm/Warm setting; electric water heater; 8 loads/week.
(4) 36¢/load as per the clothes dryers page, 8 loads a week. (Gas dryer isn't much better @ 34¢/load.)
(6) Ten 15-watt fluorescent bulbs vs. 60-watt incandescent bulbs, each burning 5.5 hours a day.
(7) Computer system sleeps for 21 hrs/day @ 5 watts, vs. on for 24 hrs/day @ 100 watts
(8) All loads washed on Warm/Warm setting. 8 loads/week. Water heated electrically. Includes water costs.
(9) Replacing a 900 kWh/year top-freezer model with a 450 kWh/year top-freezer model
(10) Used ten hours a day, five days a week. 120 watts vs. 40 watts

4.2

ELECTRICAL SAVING TIPS

Appliances & Electronics

If you live in a typical U.S. home, your appliances and home electronics are responsible for about 20 percent of your energy bills. These appliances and electronics include everything from clothes washers and dryers, to computers, to water heaters. By shopping for appliances with the ENERGY STAR® label and turning off appliances when they're not in use, you can achieve real savings in your monthly energy bill.

- Many appliances continue to draw a small amount of power when they are switched off. These "phantom" loads occur in most appliances that use electricity, such as VCRs, televisions, stereos, computers, and kitchen appliances. In the average home, 75% of the electricity used to power home electronics is consumed while the products are turned off. This can be avoided by unplugging the appliance or using a power strip and using the switch on the power strip to cut all power to the appliance.
- Air dry dishes instead of using your dishwasher's drying cycle.

- Always look for the ENERGY STAR® and EnergyGuide labels when shopping for home appliances. The ENERGY STAR® label is the government's seal of energy efficiency. The EnergyGuide label estimates an appliance's energy consumption.
- Clean the lint filter in the dryer after every load to improve air circulation.
- Consider air-drying clothes on clothes lines or drying racks. Air-drying is recommended by clothing manufacturers for some fabrics.
- Consider buying a laptop for your next computer upgrade; they use much less energy than desktop computers.
- Don't over-dry your clothes. If your machine has a moisture sensor, use it.
- Dry towels and heavier cottons in a separate load from lighter-weight clothes.
- ENERGY STAR® computers and monitors save energy only when the power management features are activated, so make sure power management is activated on your computer.
- Look for the ENERGY STAR® label on home appliances, electronics and other products. ENERGY STAR® products meet strict efficiency guidelines set by the U.S. Environmental Protection

Agency and the U.S. Department of Energy.

- For older appliances, use a power controlling device to reduce the energy consumption of the appliance's electric motor.
- Periodically inspect your dryer vent to ensure it is not blocked. This will save energy and may prevent a fire. Manufacturers recommend using rigid venting material, not plastic vents that may collapse and cause blockages.
- Plug home electronics, such as TVs and DVD players, into power strips; turn the power strips off when the equipment is not in use (TVs and DVDs in standby mode still use several watts of power).
- Saving energy starts with being an informed consumer.
- Studies have shown that using rechargeable batteries for products like cordless phones and PDAs is more cost effective than throwaway batteries. If you must use throwaways, check with your trash removal company about safe disposal options.
- There is a common misconception that screen savers reduce energy use by monitors; they do not. Automatic switching to sleep mode or manually turning monitors off is always the better energy-saving strategy.

- To maximize savings with a laptop, put the AC adapter on a power strip that can be turned off (or will turn off automatically); the transformer in the AC adapter draws power continuously, even when the laptop is not plugged into the adapter.
- Turn off your computer and monitor when not in use.
- Unplug battery chargers when the batteries are fully charged or the chargers are not in use.
- Use the cool-down cycle to allow the clothes to finish drying with the residual heat in the dryer.
- Wash and dry full loads. If you are washing a small load, use the appropriate water-level setting.
- When shopping for a new clothes dryer, look for one with a moisture sensor that automatically shuts off the machine when your clothes are dry. Not only will this save energy, it will save wear and tear on your clothes caused by over-drying.
- Turn off your personal computer when you're away from your PC for 20 minutes or more, and both the CPU and the monitor if you will be away for two hours or more.

Lighting & Daylighting

The quantity and quality of light around us determine how well we see, work, and play. Light affects our health, safety, morale, comfort, and productivity. In your home, you can save energy while still maintaining good light quantity and quality.

- Consider using high-intensity discharge (also called HID) or low-pressure sodium lights.
- Exterior lighting is one of the best places to use CFLs because of their long life. If you live in a cold climate, be sure to buy a lamp with a cold weather ballast since standard CFLs may not work well below 40°F.
- Turn off decorative outdoor natural gas lamps; just eight such lamps burning year-round use as much natural gas as it takes to heat an average-size home during an entire winter.
- Use outdoor lights with a photocell unit or a motion sensor so they will turn on only at night or when someone is present. A combined photocell and motion sensor will increase your energy savings even more.
- Consider using 4-watt minifluorescent or electro-luminescent night lights. Both lights are much more efficient than their incandescent counterparts. The luminescent lights are cool to the touch.

- If you have torchiere fixtures with halogen lamps, consider replacing them with compact fluorescent torchieres. Compact fluorescent torchieres use 60% to 80% less energy, can produce more light (lumens), and do not get as hot as the halogen torchieres. Halogen torchieres are a fire risk because of the high temperature of the halogen bulb.
- Take advantage of daylight by using light-colored, loose-weave curtains on your windows to allow daylight to penetrate the room while preserving privacy. Also, decorate with lighter colors that reflect daylight.
- Recessed downlights (also called recessed cans) are now available that are rated for contact with insulation (IC rated), are designed specifically for pin-based CFLs, and can be used in retrofits or new construction.
- Use CFLs in all the portable table and floor lamps in your home. Consider carefully the size and fit of these systems when you select them. Some home fixtures may not accommodate some of the larger CFLs.
- Consider using 4-watt minifluorescent or electro-luminescent night lights. Both lights are much more efficient than their incandescent counterparts. The luminescent lights are cool to the touch.

- Use 4-foot fluorescent fixtures with reflective backing and electronic ballasts for your workroom, garage, and laundry areas.
- Consider three-way lamps; they make it easier to keep lighting levels low when brighter light is not necessary.
- Use task lighting; instead of brightly lighting an entire room, focus the light where you need it. For example, use fluorescent under-cabinet lighting for kitchen sinks and countertops under cabinets.
- Turn off the lights in any room you're not using, or consider installing timers, photo cells, or occupancy sensors to reduce the amount of time your lights are on.
- Install task lighting – such as under-counter kitchen lights or bathroom mirror lights – to reduce the need for ambient lighting of large spaces.
- Use dimmers, motion sensors, or occupancy sensors to automatically turn on or off lighting as needed and prevent energy waste.
- Install fluorescent light fixtures for all ceiling- and wall-mounted fixtures that will be on for more than 2 hours each day.
- Use ENERGY STAR® labeled lighting fixtures.

- Consider light wall colors to minimize the need for artificial lighting.
- Use compact fluorescent light bulbs (CFLs) in place of comparable incandescent bulbs to save about 50 percent on your lighting costs. CFLs use only one-fourth the energy and last up to 10 times longer.
- Turn your lights off when you leave a room. Standard, incandescent light bulbs should be turned off whenever they are not needed. Fluorescent lights should be turned off whenever you'll be away for 15 minutes or more.
- During winter, open curtains on your south-facing windows during the day to allow sunlight to naturally heat your home, and close them at night to reduce the chill you may feel from cold windows
- Installing a skylight can provide your home with daylighting and warmth. When properly selected and installed, an energy-efficient skylight can help minimize your heating, cooling, and lighting costs.

4.3
GENERAL MONEY SAVING TIPS

Reduce Your Carbon Footprint—Use Energy Wisely

Attic
- **Door or Hatch**
  Weather-strip or insulate your attic door or hatch to prevent air from escaping from the top of your house.
- **Insulation**
  Check current insulation levels, and properly insulate a new or existing home. Insulate ceilings, walls, and floors over unconditioned crawl spaces.
- **Vents**
  Attics must be ventilated to relieve heat buildup caused by the sun. If necessary, improve attic airflow by adding or enlarging vents.

Basement
- **Heating Unit**
  As much as half of your household energy use goes to heating and cooling. Replacing older equipment with more efficient equipment will help reduce your carbon footprint and your energy costs.

Tune up your heating system in the fall to make sure that it will operate at maximum efficiency during the cold winter.

- **Air Conditioning Unit**
  Check and clean or replace air filters every month. Clean the outside condenser coil once a year.

  Schedule periodic maintenance of cooling equipment by a licensed service representative. A "tune up" in the spring will help the air conditioner run at maximum efficiency during the hot weather.

- **Water Heater**
  Decrease your carbon footprint and reduce your water heating bill by 10 percent by lowering the water heater temperature from 140°F to 120F°. (Keep the temperature at 140°F if you use an older dishwasher without a temperature booster.)

  Once a year, drain a bucket of water from the bottom of the water heater tank. This gets rid of sediment, which can waste energy by "blocking" the water in the tank from the heating element.

Locate water heaters as close to the points of hot water usage as possible. The longer the supply pipe, the more heat that is lost.

Insulate your hot water supply pipes to reduce heat loss. (Hardware stores sell pipe insulation kits.)

For older water heaters, consider buying a water heater insulation kit, which reduces the amount of heat lost through the walls of the tank.

Bathroom

- **Sink**
  To conserve water, use sink stoppers instead of letting water run while shaving.
- **Vanity Lights**
  Bathroom vanity lights are one of the most used fixtures in the average home. Use energy-efficient lighting, which can provide bright, warm light while using less energy and generating less heat than standard bulbs.
- **Shower**
  Taking an 8-minute shower every day can indirectly create as much as 1,368 pounds of CO_2 each year. By reducing your shower time to 6 minutes, you can eliminate 342 pounds of CO_2 from your

annual total.

Install a new low-flow shower head to help you conserve water and save energy—and save more than $75 each year on energy costs.

- **Toilet**
  A leaky toilet can waste 200 gallons of water per day. Be sure to repair all toilet and faucet leaks promptly.
- **Vent Fan**
  An ENERGY STAR® qualified ventilation fan will control moisture in the air while saving energy. These fans are much quieter than standard models. Fans with efficient lighting and fan motors use 65 percent less energy on average than standard models, saving $120 in electricity costs over the life of the fan.

Bedroom
- **Humidifier**
  In the winter, the air is normally dry inside your house, which is a disadvantage because people typically require a higher temperature to be comfortable than they would in a humid environment. Therefore, efficient humidifiers are a good investment for energy conservation.

- **Lighting**
  Provide task lighting over desks, tool benches, etc., so that activities can be carried on without illuminating entire rooms. Replace incandescent bulbs with energy-efficient lighting.
- **Outlets**
  Unplug any battery chargers or power adapters when electronics are fully charged or disconnected from the charger.
- **Cordless Phone**
  ENERGY STAR® qualified cordless phones that feature switch-mode power supplies and "smart" chargers will reduce your carbon footprint and add to your energy savings.

Dining Room
- **Light Switch**
  Remember to always turn off the lights when leaving a room. Turning off just one 60-watt incandescent bulb that would otherwise be on for eight hours a day can save about $15 per year.
- **Thermostat**
  Install a programmable thermostat to automatically adjust your home's temperature when you're away or sleeping.
- **Heating**
  Locate the heating thermostat on an

inside wall and away from windows and doors. Cold drafts will cause the thermostat to keep the system running even when the rest of the house is warm enough. Set the thermostat as low as comfort permits. Each degree over 68° F can add 2-3 percent to the amount of energy needed for heating.

- **Air conditioner**
  Set your thermostat to 78° F, or as high as comfort permits. When the weather is mild, turn off the AC and open the windows.
- **Vents**
  Close heating vents and radiator valves in unused rooms. Make sure that drapes, plants, or furniture do not block registers for supply or return air.

Exterior
- **Front Door**
  Install storm doors at all entrances of the house.

  Weather-strip and caulk around all entrance doors and windows to limit air leaks that could account for 15-30 percent of heating and cooling energy requirements.
- **Garage**
  Keep the overhead door of an attached garage closed to block cold winds from

infiltrating the connecting door between the house and garage.

- **Outdoor Lights**
  Install photoelectric controls or timers to make sure that outdoor lighting is turned off during the day. If using energy-efficient light bulbs, make sure that they are compatible with the controls.
- **Porch Light**
  Install energy-efficient lighting in the front porch light—one of the most-used lighting fixtures in a home. If your porch light is connected to a timer or photocell, make sure the new light bulbs are compatible with the controls.
- **Car**
  A vehicle emits 12,100 pounds of CO_2 per year on average. You can reduce your carbon footprint by combining trips and using mass transit, walking or biking when possible. Also keep your car well-maintained to maximize its fuel efficiency, safety, and reliability.

  To have a big impact, consider purchasing a hybrid car. A 4-cylinder hybrid with automatic transmission and 2-wheel drive emits nearly 40 percent less CO_2 per mile than a sports utility vehicle with automatic transmission, an 8-cylinder gasoline engine, and 2-wheel drive.

Vehicles in the United States average 231 miles per week. There are many ways to reduce your weekly mileage and shrink your carbon footprint. Try carpooling, using public transportation, and combining errands.

Keep your vehicle in good condition to maximize its efficiency. Schedule regular tune-ups, change the oil and filter every 3,000 to 5,000 miles, and make sure the tires are properly inflated.

- **Windows**
  Double-glazed windows (two panes of glass separated by a sealed air space) cut heat transfer by 40-50 percent. In extremely cold regions, triple glazing could be economically justified.

  Single-glazed windows should have storm windows. A wood or metal frame storm window provides a second thickness of glass and a layer of still air that reduces heat transmission markedly.

Kitchen

- **Dishwasher**
  Appliances account for as much as 20 percent of your energy bill. Newer, more efficient models save energy and water. If replacing your dishwasher, an

ENERGY STAR® model can reduce your carbon footprint and save more than $25 a year in energy costs.

- **Sink**
  To conserve water, repair any leaky faucets promptly. Hot water leaking at a rate of one drip per second can waste up to 1,661 gallons of water in one year— and wastes up to $35 in electricity or natural gas.

- **Refrigerator/Freezer**
  If your current refrigerator was made before 1993, it uses twice as much energy as an ENERGY STAR® model. A 1992 top-freezer model with 19-21 cubic feet indirectly emits as much as 754 pounds of CO_2 per year. A 2002 side-by-side model with 19-21 cubic feet indirectly emits as much as 442 pounds of CO_2 per year.

  Replacing an older model with a new ENERGY STAR® model can eliminate hundreds of pounds of CO_2 each year and save $45-$65 per year on your electric bills.

  Other tips:
  Keep your refrigerator at 37°- 40° F and your freezer at 5° F.

  Vacuum the condenser coils (underneath

or behind the unit) every three months.

Check the condition of door gaskets by placing a dollar bill against the frame and closing the door. If the bill can be pulled out with a very gentle tug, the door should be adjusted or the gasket replaced.

Do not put uncovered liquids in the refrigerator. The liquids give off vapors that add to the compressor workload.

- **Microwave**
  Use your microwave oven whenever possible. It draws less than half the power of its conventional oven counterpart and cooks for a much shorter amount of time.
- **Range/Oven**
  Only use pots and pans with flat bottoms on the stove. Use the right-sized pot on stove burners. A six-inch pot on an eight-inch burner wastes more than 40 percent of the burner's heat.

Develop the habit of "lids-on" cooking to permit lower temperature settings. Keep reflector pans beneath stovetop heating elements bright and clean.

Begin cooking on highest heat until liquid begins to boil. Then lower the heat control settings and allow food to simmer until fully cooked.

Cook as much of the meal in the oven at one time as possible. Variations of 25°F still produce good results and save energy.

Rearrange oven shelves before turning your oven on-and don't peek at food in the oven! Every time you open the oven door, 25° to 50°F is lost.

- **Trash**
  Recycle your newspapers, plastic and glass containers, and paper products. By cutting the amount of waste you produce in half, and doubling the amount of recycling in your household, you can save about 1,200 pounds of CO_2 per year.

Laundry Room
- **Clothes Dryer**
  Using your dryer 10 times a week indirectly creates more than 800 pounds of CO_2. Use the moisture sensor option so that the dryer turns off automatically when clothes are dry. This can help you reduce indirect dryer CO_2 emissions by 15 percent. Or, dry your clothes on a clothesline outside.
- **Clothes Washer**
  Follow detergent instructions carefully.

Adding too much detergent actually hampers effective washing action and may require more energy in the form of extra rinses.

Wash only full loads of laundry. Wash clothes in cold water. Sort laundry and schedule washes so that a complete job can be done with a few cycles of the machine carrying its full capacity, rather than a greater number of cycles with light loads.

If you're looking to buy a new washing machine, consider using a front-loading or horizontal axis machine. These new units use 30 percent less water and 50 percent less energy to make hot water and wash clothes than regular washing machines.

Living Room
- **Ceiling Fan**
  In the winter: If your ceiling fan has a switch that allows you to reverse the motor, you can operate the fan at a low speed in the opposite direction. This produces a gentle updraft, forcing warm air near the ceiling down into the living space.

  In the summer: Run the blades counterclockwise (downward) to cool more efficiently. Turning up the thermostat by just two degrees and using your ceiling fan can lower AC costs by up to 4-6 percent over the course of the cooling season. Don't forget to turn the ceiling fan off when you leave the room.

- **Fireplace**
  Make sure your fireplace has tightly fitting dampers that can be closed when the fireplace is not in use. Seal hidden air leaks in your chimney. If you have a gas fireplace, turn off the pilot light when not in use.

- **Lamps**
  Put lamps in corners of rooms where they can reflect light from two wall surfaces instead of one. Use compact fluorescent bulbs in fixtures that are on for more than two hours a day. Compact fluorescent bulbs use up to 75 percent less electricity. They also last about 10 times longer.

- **Entertainment Center**
  The average home uses 25 electronic products, accounting for up to 15 percent of household electricity use. TVs, DVD players, video games, and cable boxes can create more than 1,600 pounds of CO_2each year. Turning these products off when they're not being used can save 240 pounds of CO_2.

Better still, switching to electronic equipment with the ENERGY STAR® label will help save additional energy even when the device is turned off.

- **Windows**
  In warm weather, close your blinds and curtains during the hottest part of the day. During cold weather, keep curtains open during the daylight hours to take advantage of the sun's warmth.

Office

- **Computer and Monitor**
  Computers indirectly create nearly 500 pounds of CO_2 per year. Turning them off when not in use will save 43 pounds.

  Do not use a screen saver when your computer monitor is active. Instead, let it switch to sleep mode or turn the monitor off.

- **Printer, Fax, Copier**
  Save energy and space with a multi-function device that combines several capabilities-such as print, fax, copy, and scan. Enable power management features for additional savings. Turn off machines when not in use.

  Set office equipment to automatically switch to sleep mode. This will help equipment to save energy, to run cooler, and to last longer.

When purchasing new home office products, look for the ENERGY STAR® label to save energy.

- **Power Strip**
  Use a power strip as a central "turn off" point when you are finished using equipment. This will help eliminate the standby power consumption used by office equipment even when it is turned off.

No-Cost and Low-Cost Tips to Save Energy This Winter

Here you'll find strategies to help you save energy during the cold winter months. Some of the tips below are free and can be used on a daily basis to increase your savings; others are simple and inexpensive actions you can take to ensure maximum savings through the winter.

If you haven't already, conduct an energy audit to find out where you can save the most, and consider making a larger investment for long-term energy savings.

Take Advantage of Heat from the Sun

- Open curtains on your south-facing windows during the day to allow sunlight to naturally heat your home, and close them at night to reduce the chill you may feel from cold windows.

Cover Drafty Windows

- Use a heavy-duty, clear plastic sheet on a frame or tape clear plastic film to the inside of your window frames during the cold winter months. Make sure the plastic is sealed tightly to the frame to help reduce infiltration.
- Install tight-fitting, insulating drapes or shades on windows that feel drafty after weatherizing.
    - Find out about other window treatments and coverings that can improve energy efficiency.

Adjust the Temperature

- When you are home and awake, set your thermostat as low as is comfortable.
- When you are asleep or out of the house, turn your thermostat back 10°–15° for eight hours and save around 10% a year on your heating and cooling bills. A programmable thermostat can make it easy to set back your temperature.
    - Find out how to operate your thermostat for maximum energy savings.
    - Also see ENERGY STAR's June 5, 2008, podcast for video

instructions on operating your programmable thermostat.

Find and Seal Leaks

- Seal the air leaks around utility cut-through for pipes ("plumbing penetrations"), gaps around chimneys and recessed lights in insulated ceilings, and unfinished spaces behind cupboards and closets.
  - o Find out how to detect air leaks.
  - o Learn more about air sealing new and existing homes.
- Add caulk or weather-stripping to seal air leaks around leaky doors and windows.
  - o Find how to select and apply the appropriate caulk.
  - o Learn how to select and apply weather-stripping.

Maintain Your Heating Systems

- Schedule service for your heating system.
  - o Find out what maintenance is required to keep your heating system operating efficiently.
- Furnaces: Replace your furnace filter once a month or as needed.
  - o Find out more about maintaining your furnace or boiler.

- Wood- and Pellet-Burning Heaters: Clean the flue vent regularly and clean the inside of the appliance with a wire brush periodically to ensure that your home is heated efficiently.
  - Find other maintenance recommendations for wood- and pellet-burning appliances.

Reduce Heat Loss from the Fireplace

- Keep your fireplace damper closed unless a fire is going. Keeping the damper open is like keeping a window wide open during the winter; it allows warm air to go right up the chimney.
- When you use the fireplace, reduce heat loss by opening dampers in the bottom of the firebox (if provided) or open the nearest window slightly—approximately 1 inch—and close doors leading into the room. Lower the thermostat setting to between 50° and 55°F.
- If you never use your fireplace, plug and seal the chimney flue.
- If you do use the fireplace, install tempered glass doors and a heat-air exchange system that blows warmed air back into the room.
- Check the seal on the fireplace flue damper and make it as snug as possible.

- Purchase grates made of C-shaped metal tubes to draw cool room air into the fireplace and circulate warm air back into the room.
- Add caulking around the fireplace hearth.
  - o Find out more techniques to improve your fireplace or wood-burning appliance's efficiency.
  - o Learn tips for safe and efficient fireplace installation and wood burning.

Lower Your Water Heating Costs

Water heating can account for 14%-25% of the energy consumed in your home.

- Turn down the temperature of your water heater to the warm setting (120°F). You'll not only save energy, you'll avoid scalding your hands.
  - o Find other strategies for energy-efficient water heating.

To reduce your heating costs

• Have your heating system tuned and inspected by a service professional before each heating season. Heat losses from a poorly maintained system add up over time—sometimes at a rate of 1 percent to 2 percent a year

• Clean or replace the furnace filter often during the heating season. Furnaces use less energy if they "breathe" more easily. Follow instructions in the furnace manufacturer's manual

• Keep furniture, carpeting and curtains from blocking heat registers and air-return ducts

• If radiators are located near cold outside walls, place a sheet of aluminum foil between the radiator and the wall to reflect heat back into the room

• Don't overheat your home and overwork your furnace. Use supplemental heating equipment for hard-to-heat areas

• When replacing your furnace, look for one that's at least 90 percent efficient

• While sleeping, add an extra blanket for warmth

• Close your attic, basement, garage and exterior doors to prevent cold drafts and keep in heat

• Ceiling fans set at slow speed push warm air away from the ceiling and

move it around the room without creating
a chilling breeze. This spreads
the heat more evenly and will make
you feel more comfortable

Your Heating System's Thermostat

A setback or programmable thermostat
lets you automatically turn your heat
up before you get out of bed, down
when you leave for work, up before you
return from work and down again when
you go to bed.

Installing one before the heating season
begins could save as much as 20 percent
on your heating costs and recover
your investment in the first year. Other
simple things you can do:

• Turn down the heat. You'll typically
save 1 percent to 3 percent on your
heating costs for every degree you
dial down

• Set your thermostat at 68 degrees
when you're home and at 65 degrees
when you're away for a short time.
If you're used to higher settings, dial
down 1 degree at a time until you feel
comfortable

• Lower your thermostat to 58 degrees if you're away from home five hours or more. You use much less energy to heat the house up when you return than to keep it heated while you're away

NOTE: Warmer temperatures are recommended for homes with ill or elderly persons or infants.
Cooling

Beat the summer heat and stay comfortable with these energy savers.

• Buy an air conditioner with a high energy efficiency rating (EER). It's printed on the EnergyGuide label attached to the unit. A unit with an EER of 10 will cost half as much to operate as one with an EER of 5

• Make sure your central air conditioning system is the right size for the area you want to cool

• If you have central air conditioning, clean leaves and debris from the unit. To save energy, make sure they're not too close to the compressor because they can block airflow

• Install your air conditioner in the shade. When it's in direct sunlight, it uses more energy

• Clean the filter regularly. Dusty filters make your air conditioner work harder. Check the manufacturer's manual

More ways to stay cool

• Cool air from your window air conditioner can flow into open registers. Cover or close them so cool air doesn't escape

• Don't cool unused areas. Close doors and registers to cut energy costs

• Operate your stove, oven, dishwasher and clothes dryer in the morning or evening when it's cooler outside. They add extra heat to your home and make your air conditioner work harder

• Set the air conditioning thermostat at 78 degrees during the day when you're home and higher when you're away

• Install an automatic setback or programmable thermostat that starts your air conditioner shortly before you get home

• Reduce air conditioning needs by installing an attic fan. Hot air trapped in stuffy attics sinks into rooms below, adding to your summer cooling costs

• A ceiling fan cools fast and costs less than air conditioning

Water Heater

Water heating is a typical family's third-largest energy expense, accounting for about 14 percent of utility bills.

Try these energy savers
• Take a shower instead of a bath. You'll use less hot water

• Install a low-flow aerator or flow restrictor on an existing shower head, and you'll use less water when it seems like more! Both are inexpensive and easy to install—just screw them in

• Set your water heater temperature at 120 degrees. A family of four, each showering for five minutes, uses about 700 gallons of water a week.

By lowering the thermostat, you can
cut water heating bills without sacrificing
comfort

• Save even more by setting your water
heater to "on vacation" (if your unit
has this feature) when you're away
from home more than two days

• Turn off hot water when you don't
need it. Don't let it run when you
wash or shave

• Fix defective plumbing or dripping
faucets. A single dripping hot water
faucet can waste 212 gallons of
water a month. That can increase
your water bill and your energy bill

• Keep your hot water hot by making
sure pipes in unheated areas are
insulated

• Put an insulating blanket around your
water heater. It holds heat in

• Always use cold water when it will
do the job as well as hot

• Once a year, drain the water heater
tank completely. Then turn the incoming
water on and off, alternately, for

about 20 seconds. These actions flush minerals and sediment from inside the tank and make your water heater more efficient

NOTE: Some newer models are self-cleaning. Check the manufacturer's manual.

Dishwasher

Your dishwasher uses hot water to do its job. Here are ways to save on water heating.

• Set your dishwasher at 120 degrees or "low"

NOTE: First check your manufacturer's manual to see if you can use 120 degree water.

• Wash only full loads and use the shortest cycle to get your dishes clean

• Turn off the dishwasher after the wash and rinse cycles. When dishes air dry, you'll save on heating costs. On newer models, use the heat-off setting or the energy-saver dry option

• Avoid using your dishwasher to warm plates. The extra heat will raise your energy bill

• A dishwasher will operate more efficiently
if you unclog the drain of food
particles and clean it weekly

Stove and Oven

You can cook delicious and nutritious
meals and help lower energy costs by
following these handy tips.

• Thaw foods and cut vegetables into
small pieces. They'll take less time
to cook

• Put lids on pots and pans and make
sure they're the right size for the
burners. Foods will cook faster and
use less energy

• When the pot boils over or grease
splatters, clean the reflector pans.
They'll reflect more heat when
they shine

• If the flames on your gas stove or
oven are yellow, energy is being
wasted and the burners need adjusting.

Call an appliance repair professional

When using your oven, follow these suggestions.

• Preheat the oven only when the recipe calls for it. Don't preheat if you're using the broiler

• Use glass and ceramic dishes. They hold heat better and you can lower the oven temperature 25 degrees

• Your stove or oven may not always be the best choice! Small appliances, such as crockpots and electric frying pans, and your microwave oven may bc more energy efficient

• Open the oven door to peek at food inside, and you'll lose 25 degrees to 75 degrees of heat. It's best to look through the window or wait until the food is almost done before opening the door

• Save energy by baking an extra dish or cooking entire meals in the oven at the same time

• If you have a self-cleaning oven, clean it immediately after use. Because it's already hot, it will take less energy to get to the heat-cleaning stage

• In the market for a new gas stove? Choose a model with electronic igniters instead of pilot lights for the highest efficiency

• A microwave is best for defrosting and cooking small portions; an oven is more efficient for cooking large items such as turkeys and roasts

• When your electric burners are worn out and don't work properly, they use more energy. Save by replacing them

Clothes Washer

These helpful hints can save you money, and don't cost a penny.

• Wash and rinse your clothes in cold water instead of hot to save on water heating costs. Use a cold-water detergent

• Set the water level on your washer to match the size of the load and save two ways—on water and energy

• You'll save more by waiting to wash until you have a full load

• Add the right amount of detergent.
Too many suds make your washer work
harder and use more energy

Clothes Dryer

Here are more hot ideas to help
you save.

• Fill your clothes dryer, but don't
overload it. Your clothes will dry
faster when they have room to tumble

• Overdrying wears out your clothes
and wastes energy. Stop your dryer
when the laundry is dry by
setting the timer or using the auto
dry cycle

• Your dryer's lint trap helps warm air
flow better and dries your clothes
faster. Make sure to clean it after
each load

• Dry your laundry in consecutive loads
to take advantage of a heated dryer.
Your laundry will dry faster and use
less energy

• On sunny days, hang your clothes
outdoors to dry

Refrigerator/Freezer

Your refrigerator/freezer uses more electricity than any other appliance in your kitchen. These tips can help you use less and save more.

• Avoid opening the refrigerator or freezer door to browse. Each time you do, cold air escapes and your energy costs increase

• Let hot foods cool before putting them in your refrigerator or freezer. Hot foods cause the motor to work longer and harder

• Leave room in front of your refrigerator/ freezer to allow cold air to circulate better

• Because frozen food stays cold longer than air, it's good to keep your freezer full, but not packed. You'll save energy by placing water-filled containers in empty spaces

More facts

• Running two refrigerators increases your energy bill. Plus, older refrigerators

are less efficient than new ones.
To save, get rid of the second refrigerator

• Condenser coils remove heat from
inside the unit. Make sure they're at
least two inches from the wall and
clean them twice a year

• If cold air is escaping around the door
seal, adjust or replace the seal.
To check, close the door on a dollar
bill. If it's easy to pull out, cold air is
escaping

• If you have a manual-defrost freezer,
it will work more efficiently when ice
buildup is kept to 1/4 inch or less

• Set the refrigerator thermometer at
38 degrees to 42 degrees and your
freezer at 0 degrees to 5 degrees

Other Appliances

Most homes have at least 50 household
items that use natural gas or electricity.
Look around your home for places you
can save.

• If your water pump stays on too long
after using water, have it serviced. If
it runs whenever water is turned on,

it will wear out faster and use more
energy

• To save water and energy, turn off
faucets, indoors and out, when you're
done using them

• Make sure the toilet handle doesn't
stick after flushing. It wastes water
and makes your water pump run
longer

• Be sure the thermostats on appliances
work properly. If the thermostat
sticks, the appliance stays on and
raises your energy bill

• Turn off the humidifier or dehumidifier
when they're not needed

• After your second cup of coffee,
turn off your coffeemaker and pour
the leftover coffee into an insulated
container to keep it hot

• Turn off the TV, VCR, stereo or radio
when no one is watching or listening

• If your water pipes are wrapped with
insulating electric heat tape, turn it off
when the weather warms up

• Use small appliances that plug into electrical outlets instead of rechargeable devices, such as hand-held vacuum cleaners and lawn trimmers which use more energy

• Unplug electronics when not in use. Computers, VCRs, televisions and other electronics use energy when they're plugged in—even though they're turned off

Lighting

Follow these bright ideas to save energy.

• Choose light bulbs carefully. New compact fluorescent bulbs screw in the same as regular (incandescent) bulbs, but use only about onefourth to one-third the energy and last longer. For example, a 20-watt fluorescent bulb is equal to a 60-watt incandescent bulb. They last six times longer than regular bulbs. They can also be used as porch lights

• Install dimmer switches and threeway bulbs. They use less energy and let you enjoy a choice of lighting levels for different tasks

NOTE: Compact fluorescent bulbs can't be used with dimmer switches.

• For outdoor use, consider high-pressure sodium bulbs, which are more efficient and last longer than their incandescent counterparts

• When buying bulbs, check the lumens. The higher the lumens, the more light you'll get

More energy-saving ideas

• More light shines through when you keep dust off your lampshades, light fixtures and bulbs

• Because light bounces off walls and ceilings, you'll get more light for the money if you paint your walls light colors

• Increase the power of reflection by putting lamps in corners where two walls reflect light into the room

• Take advantage of free light from the sun by putting furniture near windows

• Place security lights on a timer or photoelectric control so they'll turn on and off automatically. Mercury vapor or high-pressure sodium lights are the best energy buys for outdoors

Waterbed

A waterbed with a heater adds to your energy bill every month. Try these tips to save.

• A solid-state heater warms your waterbed more accurately. Some newer waterbeds don't need heaters. Ask your dealer for information

• To save energy and stop heat loss, add an inch of foam around the edges and bottom of your water mattress or add a thermal liner or cover that encloses the entire water mattress

• Turn the waterbed thermostat to 90 degrees to 92 degrees in winter and down to 80 degrees to 82 degrees in summer

• Place a comforter on the bed and make your bed when you get up to keep heat in

Windows

You can stop heat from going out your windows. Take a look at these energy saving ideas.

• Drapes can cut heat loss in half if they have an insulating liner

• Let your drapes hang loose, and be sure they don't block heat registers and air-return ducts

• Vinyl shades and quilted curtains help cut heat loss. Shutters and blinds don't work as well because air travels through their open spaces

• On cool days, let the sun shine in by opening curtains, drapes, shades, shutters and blinds on the southern and eastern windows. Close them on cloudy days and at night to keep heat from escaping

• Close drapes on north-facing windows to keep the chill out in winter

• On hot summer days, open windows and doors in early morning and in the evening to let cool air in

Cut your costs

• Cut your heating losses by installing storm windows. Double-pane or triple-pane windows are best

• Replace old windows with new high-performance windows

• Repair open spaces in broken or cracked windows and door glass

• Use clear plastic or vinyl sheeting on the inside of your windows to make a temporary double-pane window. Use weatherproof tape or duct tape, trim or tacking strips to hold it in place

Insulation

More than 50 percent of energy used for winter heating leaves homes through uninsulated walls, floors, ceilings and attics. Insulation traps small pockets of air between warm and cold areas inside your home and helps keep warm air in during winter. Insulation is the key to big energy savings. Try these tips.

• Check your home's insulation. Insulation is judged by its R-value. The higher the R-value, the better the

material keeps heat in during cold
weather. Older homes should have an
insulating value of R-11 in the outside
walls and floors over unheated areas.
They should have at least an R-19
value in the ceiling or attic

• Adding batts of fiberglass insulation
in your attic is one of the most cost-effective
savings measures and one
you can do yourself

• Heat rises, but it also sinks into the
basement and crawl space through
uninsulated floors. Make your home
more comfortable and cut heating
losses by insulating floors

• Add extra insulation to floors by
covering them with a pad and rug

• Prevent heat loss as warm air travels
through heat ducts from your furnace
by wrapping heat ducts with insulation.
Also, use duct foil tape where
rectangular heat ducts join, and waterbase
acrylic latex caulk where round
and rectangular duct fittings meet

• Seal cracks where pipes, electrical
wires, vents and ducts enter your
home

• About 2 percent of air escapes your home through electrical outlets, especially on outside walls. Install insulation made for electrical outlets. You can also use safety outlet plugs to stop cold air from entering your home

• Insulate hot water pipes in unheated areas to keep hot water hot

Caulk and Weather Stripping

You'll live more comfortably when you get rid of cracks and leaks that let warm air escape from your home on cold days. Here's how.

• Seal cracks in your basement floor to keep heat in and cold air out

• Caulk windows, doors and anywhere air leaks in or out

• Weather-strip around windows and doors

• Seal cracks where pipes, electrical wires and ducts enter your home

• Seal openings where doors and

windows close into their frames with weather stripping—pieces of felt, rubber, metal or plastic that compress when you shut them

• Replace torn or worn weather stripping and caulk

NOTE: Do not caulk around your natural gas water heater exhaust pipes or furnace exhaust pipes.

Fireplace

Add to the warmth and enjoyment of a fireplace by following these tips.

• Close the damper when the fireplace isn't being used. About 14 percent of air escapes your home through the fireplace chimney

• Try not to run the fireplace and central heating system at the same time

• Seal unused fireplaces to keep heat from escaping and cold air from coming in

Pool and Hot Tub

Efficient ways to heat your pool
and hot tub can yield extra savings.
Consider these recommendations.

• Use a solar cover to get free heat
from the sun and prevent evaporation.
If too much water evaporates, the
water temperature drops

• Keep the filters clean. You'll save
energy

• Be sure the water temperature is
comfortablc, about 80 degrees.
Overheating wastes energy

• Cover your hot tub when it's not in
use to retain heat

Shopping Tips for New Appliances

When shopping for new appliances,
check EnergyGuide labels. They
provide the annual operating cost and
efficiency ratings of the appliance. Buy
the most energy-efficient model you
can to keep your energy costs down.

Building a New Home

If you're planning to build or buy a new home, be sure to add energy-saving features that can save you money year after year.

Don't forget nature's home and comfort plan. A few well-placed trees and shrubs will protect your home from winter's icy winds and summer's hot sun. Plant evergreen trees and shrubs to the north and northwest of your home.

• Plant deciduous trees (that have leaves) with high, spreading crowns to the south and west to let in the winter sun

Here are some additional energy-savers to consider for a new home.

• Install a vapor barrier facing the inside of your home to prevent damp air from getting into the insulation. It saves energy and protects the wood in your walls and attic against mildew

• Make sure the heating and cooling equipment is the right size for the area you want to heat. Oversized equipment is less efficient and more costly

• Install the water heater close to the point of use. Water stays hotter when it's piped a short distance. In large homes, two water heaters may be more efficient than one

• The larger the window area, the greater the heat loss. All window and glass areas should have storm windows or have double or triple glazing

• Insist on high energy-efficient appliances

Heating & Cooling

Heating and cooling account for about 56% of the energy use in a typical U.S. home, making it the largest energy expense for most homes. A wide variety of technologies are available for heating and cooling your home, and they achieve a wide range of efficiencies in converting their energy sources into useful heat or cool air for your home.

When looking for ways to save energy in your home, be sure to think about not only improving your existing heating and cooling system, but also consider the energy efficiency of the supporting equipment and the possibility of either adding supplementary sources of heating or cooling or simply replacing your system altogether.

- Your contractor should be able to give you energy fact sheets for different types, models, and designs to help you compare energy usage. For furnaces, look for high Annual Fuel Utilization Efficiency (AFUE) ratings. The national minimum is 78% AFUE, but there are ENERGY STAR® models on the market that exceed 90% AFUE.
- Place heat-resistant radiator reflectors between exterior walls and the radiators.
- Bleed trapped air from hot-water radiators once or twice a season; if in doubt about how to perform this task, call a professional.
- Clean warm-air registers, baseboard heaters, and radiators as needed; make sure they're not blocked by furniture, carpeting, or drapes.
- Clean or replace filters on furnaces once a month or as needed.
- Use fans during the summer to create a wind chill effect that will make your home more comfortable. If you use air

conditioning, a ceiling fan will allow you to raise the thermostat setting about 4°F with no reduction in comfort.

- Turn off kitchen, bath, and other ventilating fans within 20 minutes after you are done cooking or bathing to retain heated air.
- Install a programmable thermostat that can be adjust the temperature according to your schedule.
- ENERGY STAR® labeled products can cut your energy bills by up to 30 percent. Find retailers near you at http://www.energystar.gov/ when you're ready to replace your heating and cooling systems – as well as appliances, lighting, windows, office equipment, and home electronics.
- Insulate your hot water heater and hot water pipes to prevent heatloss.
- Insulate heating ducts in unheated areas such as attics and crawlspaces and keep them in good repair to prevent heat loss of up to 60 percent at the registers.
- Heating can account for almost half of the average family's winter energy bill. Make sure your furnace or heat pump receives professional maintenance each year. Look for the ENERGY STAR® label when replacing your system.

- Explore ways to save energy and improve the environment by taking simple steps around your home.

Summer

This summer, save money and stay cool. Keep your energy bill and your pollution output low this summer by taking a whole-house approach to cooling.

- In warm climates, where summertime heat gain is the main concern, look for windows with double glazing and spectrally selective coatings that reduce heat gain.
- If your air conditioner is old, consider purchasing a new, energy-efficient model. You could save up to 50% on your utility bill for cooling. Look for the ENERGY STAR® and EnergyGuide labels.
- Keep in mind that insulation and sealing air leaks will help your energy performance in the summertime by keeping the cool air inside.
- Plant trees or shrubs to shade air conditioning units but not to block the airflow. Place your room air conditioner on the north side of the house. A unit operating in the shade uses as much as

10% less electricity than the same one operating in the sun.

- Don't place lamps or TV sets near your air-conditioning thermostat. The thermostat senses heat from these appliances, which can cause the air conditioner to run longer than necessary.
- Consider using an interior fan in conjunction with your window air conditioner to spread the cooled air more effectively through your home without greatly increasing your power use.
- Don't set your thermostat at a colder setting than normal when you turn on your air conditioner. It will not cool your home any faster and could result in excessive cooling and, therefore, unnecessary expense.
- Set your thermostat as high as comfortably possible in the summer. The less difference between the indoor and outdoor temperatures, the lower your overall cooling bill will be.
- Whole-house fans help cool your home by pulling cool air through the house and exhausting warm air through the attic. They are effective when operated at night and when the outside air is cooler than the inside.
- For air conditioners, look for a high Seasonal Energy Efficiency Ratio

(SEER). The current minimum is 13 SEER for central air conditioners.

- During the cooling season, keep the window coverings closed during the day to prevent solar gain.

4.4
WINTER TIPS

Dial down

- The best way to manage your energy costs is by using energy wisely. For every degree you lower your thermostat, you can save about three percent on your heating bill.

- One way to accomplish this is with an automatic set-back or programmable thermostat, which can automatically lower the temperature when you're away from the house, and automatically increase it before you get home.

- Homes with proper humidity levels will provide greater comfort at lower temperatures. When humidity is kept at a proper level – about 35 percent at 70 degrees – windows will not sweat and the air won't feel dry. Replace the humidifier pad or clean it of calcium deposits for best results.

Fine tune your furnace

- Schedule a heating system check-up. A qualified heating contractor will make sure your heating system operates efficiently and delivers the maximum energy savings.

- Clean or replace your furnace's air filters as needed during the winter season. Dirty filters block the warm airflow in the home, which causes the furnace to work harder and less economically.

- Clean and vacuum ducts, vents and heat registers. Check heat registers to ensure that drapes or furniture do not block airflow.

Add insulation

- Install attic and basement insulation to keep out drafts and make your home more energy efficient and warm. Adding blown cellulose on top of rolled fiberglass insulation will increase the insulation value of your attic.

- Insulate all heating ducts located in attics and unheated crawlspaces, and make sure there are no leaks in your ductwork.

- Place an insulation blanket around your hot water heater. Electric water heaters should be placed on an insulated surface, such as foam. Adjust water heater temperature to the warm setting (about 120 degrees F).

Keep the cold air out and the warm air in

- Seal windows and doors with caulk and weather stripping to block unwanted drafts. If caulk cracks and peels away, it allows your home's heat to escape.

- Install a fireplace door to prevent cold air from entering your home. And, be sure to close the damper unless a fire is burning. Keeping the damper open is like having a window wide open during the winter. However, if you have a gas fireplace, the flue should be partially open to allow fumes from the pilot light to escape.

- Use kitchen, bath and other ventilating fans wisely. In just one hour, these fans can pull out a houseful of warm air. Turn fans off as soon as they have done their job.

Let the sun shine in

- Open curtains on south facing windows during the day to allow sunlight to naturally heat your home. Be sure to close the curtains at night to reduce the chill you may feel from cold windows.

Winter

This winter, save money and stay warm. Keep your energy bill and your pollution output low this winter by taking a whole-house approach to heating.

- During the heating season, keep the draperies and shades on your south facing windows open during the day to allow the sunlight to enter your home and closed at night to reduce the chill you may feel from cold windows.
- Set your thermostat as low as is comfortable when home.
- By resetting your programmable thermostat from 72 degrees to 65 degrees for eight hours a day (for instance, while no one is home or while everyone is tucked in bed) you can cut your heating bill by up to 10 percent.
- Weatherize your home—caulk and weatherstrip any doors and windows that leak air.

- Properly maintain and clean heating equipment.
- Replace furnace filters regularly.
- Check the insulation in your attic, ceilings, exterior and basement walls, floors, and crawl spaces to see if it meets the levels recommended for your area.

GLOSSARY

Lighting Principles and Terms

To choose the best energy-efficient lighting options for your home, you should understand basic lighting principles and terms.

Light Quantity

Illumination

The distribution of light on a horizontal surface. The purpose of all lighting is to produce illumination.

Lumen

A measurement of light emitted by a lamp. As reference, a 100-watt incandescent lamp emits about 1750 lumens.

Footcandle

A measurement of the intensity of illumination. A footcandle is the illumination produced by one lumen distributed over a 1-square-foot area. For most home and office work, 30-50 footcandles of illumination is sufficient. For detailed work, 200 footcandles of illumination or more allows more accuracy and less eyestrain. For simply finding one's way around at night, 5-20 footcandles may be sufficient.

Energy Consumption

Efficacy
The ratio of light produced to energy consumed. It's measured as the number of lumens produced divided by the rate of electricity consumption (lumens per watt).

Light Quality

Color temperature
The color of the light source. By convention, yellow-red colors (like the flames of a fire) are considered warm, and blue-green colors (like light from an overcast sky) are considered cool. Color temperature is measured in Kelvin (K) temperature. Confusingly, higher Kelvin temperatures (3600-5500 K) are what we consider cool and lower color temperatures (2700-3000 K) are considered warm. Cool light is preferred for visual tasks because it produces higher contrast than warm light. Warm light is preferred for living spaces because it is more flattering to skin tones and clothing. A color temperature of 2700-3600 K is generally recommended for most indoor general and task lighting applications.

Color rendition
How colors appear when illuminated by a light source. Color rendition is generally considered to be a more important lighting quality than

color temperature. Most objects are not a single color, but a combination of many colors. Light sources that are deficient in certain colors may change the apparent color of an object. The Color Rendition Index (CRI) is a 1-100 scale that measures a light source's ability to render colors the same way sunlight does. The top value of the CRI scale (100) is based on illumination by a 100-watt incandescent light bulb. A light source with a CRI of 80 or higher is considered acceptable for most indoor residential applications.

Glare
The excessive brightness from a direct light source that makes it difficult to see what one wishes to see. A bright object in front of a dark background usually will cause glare. Bright lights reflecting off a television or computer screen or even a printed page produces glare. Intense light sources such as bright incandescent lamps are likely to produce more direct glare than large fluorescent lamps. However, glare is primarily the result of relative placement of light sources and the objects being viewed.

Lighting Uses

Ambient lighting
Provides general illumination indoors for daily activities, and outdoors for safety and security.

Task lighting
Facilitates particular tasks that require more light than is needed for general illumination, such as under-counter kitchen lights, table lamps, or bathroom mirror lights.

Accent lighting
Draws attention to special features or enhances the aesthetic qualities of an indoor or outdoor environment.

START TODAY

There is no time to waste! Start Today!!!

Visit us on the web where you can download and print your home assessment kit and follow the easy to understand checklist to see where you are losing money in your home.

You can also order money saving products and services on our website as well!

So don't hesitate...Start Today!

The only thing you have to lose is more money for wasted energy...

American Environmental

American Environmental is not a Government Agency, but a private company whose main focus and goal is to reduce consumer energy waste to protect our Earth's delicate eco system and save consumers millions of dollars in wasted energy per year. The services we provide are available online as well as products and more. Visit us online today and start saving your money, and our eco-system.